INTEGRATED **WATER RESOURCES MANAGEMENT** IN PRACTICE

INTEGRATED **WATER RESOURCES MANAGEMENT** IN PRACTICE

Better water management for development

Edited by Roberto Lenton *and* Mike Muller

With the assistance of Sarah Carriger

earthscan
from Routledge

First published by Earthscan in the UK and USA in 2009

For a full list of publications please contact:
Earthscan
2 Park Square, Milton Park, Abingdon, Oxon OX14 4RN
711 Third Avenue, New York, NY 10017

Earthscan is an imprint of the Taylor & Francis Group, an informa business

ISBN: 978-1-84407-649-9   Hardback
        978-1-84407-650-5   Paperback

Typeset by 4word Ltd, Bristol
Cover design by Ruth Bateson
Cover images: industrial image © Jörg Hochscherf/Fotolia.com; cup of clean water © Claudia Dewald/
        iStockPhoto.com; Komati River, South Africa © Mike Muller

A catalogue record for this book is available from the British Library

Library of Congress Cataloging-in-Publication Data

Integrated water resources management in practice: better water management for development / edited by Roberto Lenton and Mike Muller.
        p. cm.
  Includes bibliographical references and index.
  ISBN 978-1-84407-649-9 (hardback) – ISBN 978-1-84407-650-5 (pbk.)
  1. Water supply–Management. 2. Water resources development. 3. Sustainable development.
  I. Lenton, R. L. II. Muller, Mike.
  TD345.I538 2009
  363.6'1–dc22
                                2008041791

Contents

Part Two – Basin Level

Part Three – National Level

List of Tables, Boxes and Figures

Tables

Boxes

Figures

Foreword

Water is an integral part of the lives and livelihoods of all of us. Our health, our food, our energy security and our environment all depend on investments in water resources management. Without good water management, we will not be able to achieve sustainable development or reduce poverty. Nor will we be able to respond effectively to emerging new challenges such as climate change adaptation.

The Global Water Partnership's (GWP) vision is for a water-secure world, in which floods and droughts are effectively managed, and the necessary quantity and quality of water is available for health, economic development and the preservation of ecosystems. Our mission is to support the sustainable development and management of water resources at all levels. Following periods of conceptualizing and advocating an integrated water resources management (IWRM) approach and of establishing locally-owned regional and country partnerships, we now are poised to take on new challenges. Our new strategy seeks to support countries to improve water resources management, putting IWRM into practice to help countries towards growth and water security.

The publication of *Integrated Water Resources Management in Practice* is therefore exceptionally timely. The book contains important lessons that will guide and inspire GWP and its partners in the years to come. It emphasizes, for example, that pragmatic, incremental approaches, which take into account contextual realities, seem to have had the greatest chance of working in practice. And it highlights the fact that integrated water resources management is not a prescription, but rather an approach that offers a practical framework within which the problems of different communities and countries can be addressed. *Integrated Water Resources Management in Practice* will therefore hopefully put to rest the concerns of some that IWRM is an unrealistic and impractical approach.

I am grateful to GWP's Technical Committee for its leadership in preparing this book. I am sure the ideas, experiences and lessons in this book will greatly contribute to advancing thinking and action on water management. I strongly recommend it to those interested in and concerned with the management of water resources for sustainable development and the reduction of poverty.

Letitia A. Obeng
Chair
Global Water Partnership

Preface

This book is about the role of water in development. It's about water and people and the diversity of areas in which better water management can make a difference to lives, livelihoods and the environment. Our aim is to help readers understand, in practical terms, the approach now known as 'integrated water resources management'. We do so by telling the stories of people and institutions around the world who have found ways to improve water management in a variety of settings.

Integrated Water Resources Management in Practice is the result of intensive effort, discussion and debate by the Technical Committee (TEC) of the Global Water Partnership over a four-year period, beginning in 2004. The development of this book benefited from many thoughtful deliberations during many TEC meetings that guided the conceptual framework, the case study selections, and the lessons learned. Participants in these discussions have included our present and former TEC colleagues Mohamed Ait-Kady, Akiça Bahri, Hartmut Brühl, Jennifer Davis, Malin Falkenmark, Simi Kamal, Uma Lele, Humberto Peña, Judith Rees, Peter Rogers, Claudia Sadoff, Miguel Solanes, Eugene Terry, Patricia Wouters, Albert Wright and Yang Xiaoliu. Other members of the GWP community who have actively participated in these discussions over the years include Leanne Burney, Sarah Carriger, Margaret Catley-Carlson, Elisa Colóm, Mercy Dikito-Wachtmeister, Emilio Gabbrielli, Nighisty Ghezae, Gabriela Grau, Björn Guterstam, Alan Hall, Torkil Jonch-Clausen, Wayne Joseph, Axel Julie, Aly Kerdany, Henrik Larsen, Suresh Prabhu, Michael Scoullos, Vadim Sokolov and

Simon Thuo. We are grateful for the contributions of all these individuals, and especially the members of the small task force that the TEC established in 2006 to oversee the preparation of this book and that included, in addition to ourselves, Hartmut Bruhl, Sarah Carriger, Simi Kamal and Judith Rees. We are grateful to them all for their inspiration, encouragement and hard work throughout the process. While all members of the Committee contributed actively to the book in a number of different ways, we as editors take full responsibility for any errors or omissions in the contents of this book.

We are also especially grateful to our chapter authors for joining us in the preparation of this book, for sharing our enthusiasm, and for showing much patience and goodwill as we commented on and edited, sometimes extensively, the original material. A few of the chapter authors were directly involved in the cases described; these authors have in the process had to subject themselves to some rigorous criticism, which they have willingly endured. The first and last chapters have been written by us as co-editors. In all chapters, monetary figures in different currencies have been converted to US$ equivalents to make the numbers more meaningful to readers in different parts of the world, using the nominal exchange rate in effect on 1 September 2008. We have provided short biographies of each of the chapter authors in the list of contributors.

Besides the individual chapter authors, we would like to thank specifically those who generously donated their time to review and critique the case study chapters, including Jørn-Ole Andreasen, Carl Bauer, Mariann Brun, Per Grønlund, Yang Guowei, Kenzo Hiroki, Masaki Hirowaki, M.R. Khurana, Andreas S. Kofoed, Klaus Kolind-Hansen, Doug Merrey, Khalid Mohtadullah, Francois Molle, Masahisa Nakamura, Madiodio Niasse, Jerry Priscoli, Atiq Rahman, Djoko Sasongko, Barbara Schreiner, Lars Schrøder, David Seckler, Richard Thomsen, Claus Vangsgaard, Flip Wester and Dong Zheren.

The book could not have been produced without the help of Sarah Clarice Carriger, Science Writer & Communications Consultant, who did much of the initial case study research and edited the final manuscript in both style and content. We are also very grateful to Christie Walkuski, who not only co-authored one of the case study chapters but also provided consistent writing, research and coordination support throughout the process. Others who played important writing and research roles included Cheryl Antao, who prepared an extensive literature review as well as some graphs and figures; Leanne Burney, who undertook some of the early case study research and proposal development; Sarah Dobsevage, who prepared some of the case study summaries in the initial stages of the book's development; Kristen Lewis, who contributed to some of the early drafts of Chapter 1; Harold Thompson, who checked all the references in great detail and secured all copyright permissions; and Kytt MacManus at the Center for International Earth Science Information Network (CIESIN) of the Earth Institute at Columbia University, who provided much-needed assistance with the production of maps and graphics. The project was also assisted by GWP's vast network of regional and country partnerships, who contributed suggestions of cases and comments throughout the production and writing process.

Many people in the Secretariat of the GWP in Stockholm contributed to the overall effort. We would especially like to thank James Lenahan, the former head of Communications, who provided much helpful publishing advice. Emilio Gabbrielli, former Executive Secretary, Martin Walshe, Acting Executive Secretary during the time the book was completed, and Steven Downey, Director of Communications, also provided much needed support, advice and encouragement. We are also grateful to Margaret Catley-Carlson, the former GWP Chair, for her advice and guidance since inception, and to Letitia A. Obeng, the current Chair of GWP, for her strong support for the endeavour and for writing the foreword to this book. We have also appreciated the excellent support received from Tim Hardwick, Gina Mance, Andrew Miller, Olivia Woodward, Alison Kuznets and Claire Lamont of the staff of Earthscan, and John Roost of 4word Ltd Page & Print Production.

Finally, we would like to extend a special thank you to Ruth Levine of the Center for Global Development and her colleague Jessica Gottlieb, for sharing with us the lessons they learned in the preparation of *Millions Saved: Proven Successes in Global Health*, which served as an inspiration for this book. (Ruth Levine and the What Works Working Group, *Millions Saved: Proven Successes in Global Health*. Washington: Center for Global Development, 2004, 180 pp. in paper cover, available at www.cdgev.org.)

Our greatest debt of gratitude goes to those whose stories we tell: the doers and thinkers in communities, local authorities, NGOs, private entities, government agencies and regional or international bodies who are bringing innovative solutions and common sense to bear on water-related problems, and who are lighting the way for others to practise better water management. We thank them all.

Roberto Lenton and Mike Muller
September 2008

Contributors

Akiça Bahri

Akiça Bahri is a member of the Global Water Partnership's Technical Committee and is Director for Africa for the International Water Management Institute. She is an agronomy engineer with PhD degrees from universities in France and Sweden. She has worked in the fields of agricultural use of marginal waters (brackish and wastewater), sewage sludge and their impacts on the environment. She has been working for the National Research Institute for Agricultural Engineering, Water and Forestry in her home country of Tunisia, where she was in charge of research management in the field of agricultural water use. She works on policy and legislative issues relating to water reuse and land application of sewage sludge, and is a member of several international scientific committees.

Boubacar Barry

Boubacar Barry has over 25 years of experience as a scientist in irrigation, hydrology, soil and water conservation engineering. He has extensive experience in planning, implementing and supervising research projects as well as development projects in West Africa. His research interests in soil and water engineering relate to basin hydrology, water quality, drainage, irrigation and erosion control, with special emphasis on rainwater harvesting. His interest in practical research application is complemented by his expertise in Remote Sensing/GIS, numerical modelling and expert systems.

Hartmut Brühl

Hartmut Brühl is a member of the Global Water Partnership's Technical Committee and is a consultant for water resources management and hydraulic engineering (ports, coasts and rivers). He has in-depth experience in project management, contract management and international cooperation. A trained civil and hydraulic engineer with MSc and PhD degrees from the University of Hanover, Germany, he has more than 39 years of experience in the international water sector. He was an Associated Professor for Hydraulic Engineering before joining an internationally focused consulting company responsible for water resources and hydraulic engineering projects mainly in West Africa, Latin America and Southeast Asia. On a UNDP assignment, Brühl worked for the Mekong Committee, Bangkok, as Senior Advisor for River Basin Management. For 12 years he worked for the KfW Banking Group, the official German development bank, as a technical expert, and later as a First Vice President and head of the Technical Department.

Mogens Dyhr-Nielsen

Mogens Dyhr-Nielsen is head of Capacity Development Networks in Denmark. Dyhr-Nielsen has 35 years of experience in management and research on natural resources and environment issues. He has worked for the Danish government, and has assisted international organizations including UNESCO, WMO and UNEP. He has been involved in integrated water resources management since the Mar del Plata

Conference in 1977, with a focus on groundwater development, nutrient pollution and ecosystem deterioration. He was director of DANCED, the Danish programme for environmental assistance to Southeast Asia and Southern Africa, and recently assisted UNEP in providing support to the ASEAN governments on establishing IWRM 2005 strategies and roadmaps.

Jorge Hidalgo

Jorge Hidalgo is a researcher at the Mexican Institute for Water Technology. He has been a professor at the National University of Mexico since 1996 and also serves as technical arbiter of the Journal Ingeniería Hidráulica en México (Hydraulic Engineering in Mexico). He is a member of the Mexican Academy of Engineering, the Mexican Hydraulics Association and the American Water Resources Association. Hidalgo has published more than 50 papers and reports on hydrology and water resources management. He received the 1980 National Public Administration Award. His recent research interests are dynamic river basin models and integrated water resources management and planning.

Deepa Joshi

Deepa Joshi works for Winrock International as Innovations Program Officer. She has extensive work experience in South Asia and has also worked in Africa and Latin America. Her skills range from project management for international and grassroots NGOs to leadership and management of international, multi-disciplinary research on poverty and gender in South Asia and Africa. She has worked with the International Water Management Institute as Researcher on gender, poverty and livelihoods and, earlier, as a Senior Fellow at the Institute of Irrigation and Development Studies, University of Southampton. She is author of several academic papers on water and development, poverty, gender and development.

Simi Kamal

Simi Kamal is a member of the Global Water Partnership's Technical Committee, the Chair of Hisaar Foundation – a Foundation for Water, Food and Livelihood Security and the Chief Executive of Raasta Development Consultants. A geographer from the University of Cambridge, she has 28 years of work experience in policy, research, evaluation, advocacy and capacity building in water, IWRM, sanitation, irrigation, drainage, agriculture and related sectors across the world. Her other areas of specialty include empowerment and advancement of women, and the building of cross-sectoral collaborative platforms, development institutions and networks. She is currently a member of Pakistan's National Commission on the Status of Women, and of several development and private sector organizations, boards and committees. She has over 450 reports, papers, articles, handbooks, modules and book chapters to her credit across many development sectors, and is a consultant to governments, UN agencies, the World Bank and ADB in Pakistan and abroad.

Roberto Lenton

Roberto Lenton is Chair of the Global Water Partnership's Technical Committee and is a Member of the Inspection Panel of the World Bank. A specialist in water resources and sustainable development, he also serves as Chair of the Water Supply and Sanitation Collaborative Council, Member of the Board of Directors of WaterAid America, and Senior Advisor to the International Research Institute for Climate and Society at Columbia University. Dr Lenton was formerly the Director of the Sustainable Energy and Environment Division of the United Nations Development Programme; Director General of the International Water Management Institute; and Program Officer in Rural Poverty and Resources at the Ford Foundation in New Delhi and New York. He has also served on the faculty and staff of Columbia University and the Massachusetts Institute of Technology. A citizen of Argentina with a Civil Engineering degree from the University of Buenos Aires and a PhD degree from MIT, Dr Lenton is a co-author of *Applied Water Resources Systems* and a lead author of *Health, Dignity and Development: What will it take?*, the final report of the United Nations Millennium Project Task Force on Water and Sanitation, which he co-chaired.

Mike Muller

Mike Muller is a member of the Global Water Partnership's Technical Committee. An engineer by training, he was the Director General of South Africa's Department of Water Affairs and Forestry from 1997 to 2005, where he led the development and implementation of new water resources and water services policies as well as water-sharing agreements and cooperation projects with Mozambique, Swaziland and Lesotho. He represented South Africa in many international water fora and was a member of the United Nations Millennium Project Task Force on Water and Sanitation. Muller also worked for nine years for the Mozambican government and has engaged extensively in broad development policy issues – his publications on health and development include *The Health of Nations, Tobacco and the Third World – Tomorrow's Epidemic* and *The Babykiller*. Currently he is a visiting professor at the Graduate School of Public and Development Management at the University of Witwatersrand, Johannesburg.

Regassa E. Namara

Regassa E. Namara is a Senior Research Economist at the International Water Management Institute. He has a PhD degree from the University of Goettingen, Germany, and started his research career with the Ethiopian Institute of Agricultural Research in 1987. His expertise includes agricultural economics, microeconomics of agricultural water management, innovation adoption and diffusion research, research and development impact evaluations, and socio-economics of rural development. Currently, he serves as guest lecturer of economics and financial analysis at the UNESCO-IHE Institute for Water Education. He has substantial field research experience in many African and Asian countries.

Humberto Peña

Humberto Peña is a member of the Global Water Partnership's Technical Committee and a hydraulic engineer with a degree from the Catholic University of Chile. As Chile's National Director of Water Resources from 1994 to 2006, he had a defining role in the National Water Policy and the Chilean Water Reform, and was responsible for application of Chile's water laws. He has worked in several national and international research programmes, and published more than 100 papers on water policies, hydrology and water resources. He has been a consultant on water policy for FAO, WMO/WB, ECLAC-UN, CARE and PNUMA projects in several Latin American countries.

Mary Renwick

Mary Renwick is a water resources specialist and economist with over 18 years' experience developing, managing and implementing water resource projects, programmes and initiatives, and has worked in over 20 countries in sub-Saharan Africa, South and Southeast Asia (and with a number of Native American tribes). She has worked with donors, governments, NGOs, local communities and in academia. Since 2005, she has served as Director of Winrock International's Water Innovation Program. Before joining Winrock, Mary was a Senior Fellow in Water Policy and Economics at the University of Minnesota and an Adjunct Professor in Applied Economics. She was also a Fulbright Visiting Professor in Interdisciplinary Water Resources Management at Khon Kaen University in Thailand. She holds a PhD degree in applied economics from Stanford University and is the author of over 25 books, peer-reviewed articles and papers.

Peter Rogers

Professor Peter Rogers is Gordon McKay Professor of Environmental Engineering in the School of Engineering and Applied Sciences at Harvard University. He is a senior advisor to the Global Water Partnership, a Fellow of the American Association for the Advancement of Science, a member of the Third World Academy of Sciences, and is a recipient of Guggenheim and Twentieth Century Fellowships and the 2008 Warren Hall Medal of the Universities Council on Water Resources. His research interests include the consequences of population on natural resources development, conflict resolution in international river basins, improved methods for managing natural resources and the environment, and the development of indices of environmental quality and sustainable development.

Amina Siddiqui

With two degrees from the University of Karachi, Amina Siddiqui's special areas of work include water, gender, environment and the health sector. She has worked on numerous research studies in the Asia Pacific region relating to water, IWRM and gender, and compiled reports and case studies, including several for the GWP Toolbox. She is one of the original members of the Women and Water Network South Asia and has continued to contribute to this network.

Christie Walkuski

Christie Walkuski has served as Administrative Coordinator for the Global Water Partnership's Technical Committee since 2003 and is Program Assistant for the International Research Institute for Climate and Society in their Asia Program. She is a MSc degree candidate in the Graduate School of Architecture, Planning and Preservation at Columbia University, focusing her research on housing and community-development aspects of urban planning in the USA. Before joining the Global Water Partnership, she worked for several non-profit advocacy and community-building organizations, working on issues of hunger and homelessness, poverty and environment. She earned a BA degree in Environmental Science from the State University of New York at Purchase.

Xiaoliu Yang

Xiaoliu Yang is a member of the Global Water Partnership's Technical Committee and is Professor of Water Resources at Peking University, China. He worked for nine years with China's central government and served as scientific advisor to the SUEZ group. His research interests include integrated management of water and land in river basins, the establishment of a water-saving society, the development of indicators of water security, and climate change vulnerability and disaster risk assessments and adaptation for urban areas in relation to water supply and sanitation.

1

Introduction

Roberto Lenton and Mike Muller

Across the world, economies are expanding, cities are spreading, many services are improving and an increasing proportion of still growing populations are enjoying better standards of living and quality of life. In rural areas, more people are growing enough food to sustain themselves as well as nearby cities.

One thing that is not growing is the natural resource base that underpins this economic and social development, of which water is a critical element. Contrary to the gloomy predictions of the doomsayers, the world does not yet face a global water crisis, nor does it need to. But if we do not develop effective approaches to address the growing stress on water resources, challenges could all too easily become crises. And we certainly do have a challenge if we are to achieve our economic and social goals *and* ensure we still have a natural resource endowment to bequeath to future generations.

So the management of water has rightly come to preoccupy many different sectors of many different societies. There is growing recognition that unless we manage our water better, we will not achieve our broader development goals. This recognition is accentuated by the looming impact of climate change. It is widely understood that our energy use is driving the changes that are increasingly being observed around us. Less well understood is the fact that higher global temperatures will impact most immediately on water resources, on both the supply and demand side. They will make rainfall more variable and more intense, while higher rates of evaporation and associated impacts on vegetation are likely to result in more extremes of river flows – contributing to both floods and droughts. Climate change will also significantly affect the requirements for water by different sectors, especially agriculture.

Water is a renewable resource so it will never 'run out'. But many societies take for granted the relatively predictable availability of the water on which their economies and environments depend. They do so at their peril. Already, large river systems in many parts of the world have been identified as being in a state of serious water stress (see Figure 1.1), meaning that withdrawing more water from these systems will result in irreversible damage to ecosystems (Smakhtin et al, 2004). And, while it cannot yet be scientifically confirmed, early indications are that, as predicted, the size and frequency of extreme events is increasing in response to rising global temperatures.

Our main goal in writing this book is thus to make the case for the better management of water resources – both today and in an increasingly uncertain tomorrow – to support growth and development and avoid the potentially catastrophic human, economic and environmental consequences of continuing with a business-as-usual approach.

Specifically, we want to help development policy-makers and practitioners in different sectors to understand the principles and practice of what is

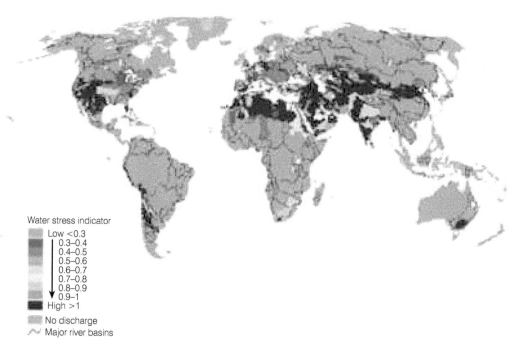

Water stress indicator

Low <0.3
0.3–0.4
0.4–0.5
0.5–0.6
0.6–0.7
0.7–0.8
0.8–0.9
0.9–1
High >1
No discharge
Major river basins

Note: The Water Stress Indicator is based on the proportion of available water withdrawn from a river system after environmental needs have been met. Basins with a water stress indicator above 0.4 are experiencing some degree of environmental stress.
Source: Smakhtin et al, 2004

Figure 1.1   Water stress in major river basins

known internationally as 'integrated water resources management', IWRM for short, so that they can use it to address their water and development challenges, particularly those intractable problems that narrower approaches have failed to resolve.

To do this, rather than describing an abstract theory, we have used case studies to illustrate the approach. The diverse experiences that we have documented present in practical terms what has been done successfully in a variety of settings to achieve a range of different goals.

The book thus demonstrates how better water management, embodying some of the key principles of IWRM, has made a positive contribution in areas as diverse as agriculture, transport, energy, industry, job creation and environmental protection. But inevitably and importantly, because we are dealing with real life,

the cases also highlight the limits to what has been achieved to date.

In the remainder of this chapter we first describe the management of water in the broader context of sustainable development, introducing the reader to the concept of water resources management and describing the key features of the IWRM approach. We then outline the book's conceptual framework, providing a brief introduction to the cases studies found in this book and the key elements of the framework, and summarize some of the key lessons that emerge from the cases.

Water and development

To begin, it is helpful to locate water resources management in the broader development context because the objective is not water management for its own sake but water management to support sustainable human development.

Water's influence on development has been recognized for centuries. In the 1770s, when Adam Smith (1776) sought to understand why some countries made economic progress while others stagnated, he concluded in his seminal *Wealth of Nations* that water was a key factor. Although his focus was on the economic advantage offered by easy navigation, it is notable that he describes the successes of regions that are still today amongst the great bread-baskets of the world, due largely to the availability of water and fertile riverine soils.

Another commentary about the influence of sound water management on social progress comes from Napoleon Bonaparte who, while in Egypt, commented that 'under a good administration the Nile gains on the desert; under a bad one the desert gains on the Nile' (cited in Moorehead, 1962: 48). And water management interventions have continued to reflect and define broader development initiatives at local, regional and national level, and sometimes beyond national borders.

But it is not so much annual water availability that has the greatest impact on the economic development of nations, but water variability – the variation of available water resources in space and time (Brown and Lall, 2006). In other words, it is a country's ability to manage water, more than the plentifulness of its water resources, that impacts economic development.

In addition to serving as a brake on economic growth, the variability of the water cycle, in particular droughts and floods, imposes huge costs on vulnerable people and national economies. A well-known recent example is the floods caused by Hurricane Katrina. The fallout from this natural disaster – which was exacerbated, it is generally agreed, by poor water management – landed the hardest on the poorest populations. However, while for a rich country like the USA, the impact of such a disaster on the overall economy is minimal (some authorities even reported a positive impact on GDP: Standard and Poors, 2005), for poor countries, droughts and floods often have devastating impacts on national economies. For example, the cost of the 1997/1998 floods in Kenya represented 11 per cent

Table 1.1 *Kenya: the economic impact of flood and drought*

1997–1998 El Niño Flood Impacts	US$ millions	% of total cost
Transport infrastructure	$777	88%
Water supply infrastructure	$45	5%
Health sector impacts	$56	6%
Total economic flood impacts	$878	

Flood impacts as % of GDP 1997–98 – 11%

1998–2000 La Niña Drought Impacts	US$ millions	% of total cost
Hydropower losses	$640	26%
Industrial production losses	$1,400	58%
Agricultural production losses	$240	10%
Livestock losses	$137	6%
Total economic drought impacts	$2,417	

Drought impacts as % of GDP 1998–2000 – 16%

Source: World Bank, 2004

of the country's GDP, and the costs associated with drought in 1999–2000 were even higher (see Table 1.1). If current climate change predictions are correct, many countries can expect an increasing number of such economic hits in the future.

Given the linkages between water management and broader development, it is not surprising that the challenge of managing water in complex societies mirrors many other management challenges and that there are often close parallels in approaches to governance and public administration. Early stages in the growth and development of great civilizations have typically been marked by the expansion of infrastructure, whether it has been the roads and aqueducts of Rome or the great wall and major water works of China. Similarly, the management of water has often been characterized by the development of infrastructure for flood protection, water storage and transport.

But as our societies and the economies that support them grow more complex, so too do the governance instruments we need to manage them, the institutions that allocate resources among different users and settle disputes, the financial mechanisms through which the necessary infrastructure and its ongoing management are funded. This has happened in water as it has happened in other sectors.

Advances in public administration have sometimes been led by developments in the water sector. It was the need to heal a septic river Thames that led to the establishment of local government institutions in London, just as the need for cooperative management of flooding and drainage in the Netherlands saw some of the earliest local administrations emerge on the European continent.

Particularly as the interdependency between water uses and users has grown, so too has the need to move beyond formal, hierarchical structures of management to institutions which engage and involve users and others affected by the resource. The current trend in decentralization is driven in part by the simple administrative logic of

subsidiarity – decentralizing activities which can best be performed at lower levels without central control. But it also reflects the understanding that, in matters of common concern, the appropriate and effective engagement of interested parties in key decisions can improve the quality of those decisions – as well as compliance with them.

The broader culture of democratic participation is mirrored in best practices in water management, whether at the level of a single village and stream or at a continental scale where the management of great rivers that traverse many countries can only be effective if all parties are drawn together to cooperate. And in water, as in other fields of endeavour, the challenge is to promote mechanisms of participation that lead to optimal outcomes rather than delaying adaptive responses until crises emerge that impose short-term, often inappropriate, responses.

Indeed, one of the challenges of promoting better water management is to do so in a manner which is compatible with broader approaches to governance and public management. This generic problem is often neglected when enthusiastic proponents of new development approaches seek to transport them from jurisdictions in Europe and North America to societies with very different systems and philosophies of government (Matas, 2001).

But all societies grapple with the challenges of managing large, complex and interconnected systems. This is exemplified by current approaches to sustainable development: the attempt to manage our resources in a manner that ensures tomorrow's generations can draw the same benefit from them as we do. The management of water resources is amongst the most challenging dimensions of sustainable development. It is these challenges that the approaches being implemented under the banner of 'integrated water resources management' address, explicitly placing environmental sustainability as one of the three key objectives of water management, along with social equity and economic efficiency.

What do we mean by water resources management?

The development of infrastructure, the allocation of the resource, the implementation of incentives for its efficient use, its protection, as well as the financing of all these activities: these are among the activities that collectively constitute water resources management.

While the management of water use – whether for domestic, agricultural or industrial purposes – is usually well understood, the management of the resource to enable the use to take place is not.

This incomplete understanding is often because the first interventions in water resources management are driven by individual users who abstract and store water for their particular purposes. But the interconnected nature of the water cycle means that individual actions often have impacts – both positive and negative – for other users.

As mentioned above, infrastructural aspects such as changing the course of rivers to support transport and provide flood protection and the construction of dams to store water are the most visible kind of water management. The visible interventions have sometimes led to an association of water resource management with the construction of large infrastructure, but this is misleading.

Just as important are those 'soft management' measures which serve to regulate the use of the resource and potential conflict between users. Where water is scarce, some mechanism has to be found to share it between users in a predictable fashion. In many societies, in Sri Lanka for example, rules governing this have emerged over many centuries of practice; in others it is codified by law or, where water flows between sovereign states, by international treaties.

Increasingly important are other obligations, such as the need to protect water from pollution so that others can use it. Regulation of water quality is often far more difficult to manage than the regulation of water quantity since it requires technical capacity to monitor as well as to enforce. In some jurisdictions – France is an immediate example – bringing water users and polluters together to discuss how best to maintain acceptable water quality has helped to ensure that the interventions made are cost effective since they are funded, in part, by those who caused the problems.

The question of who pays for what and how water management activities are to be funded is an important one. Historically, they have been funded out of the public purse since many of the benefits (whether they were security from floods or reliability of harvests) are public goods rather than immediate benefits to easily identifiable individuals. More recently it has been found that, in water as in other areas of environmental management, the imposition of environmental charges on sectors which use rivers and lakes as a sink in which to dispose of their wastes can not only raise revenue but, as important, provide economic incentives to manage wastes in a manner that has less impact on the natural environment and other users and to use water more efficiently, particularly where the intensity of resource use is increasing.

Clearly, therefore, water resources management encompasses a wide range of activities, from the development of infrastructure and the allocation of the water resource to financing arrangements and the implementation of incentives for the efficient use and protection of water.

Integrated water resources management as an element of sustainable development

The history of water management is one of emerging challenges as the number of uses and users has grown and the intensity of their water use, often measured in terms of the proportion of available water that is actually used, has increased.

While water was plentiful and abstractors few, the rules for water sharing in most societies were very basic. Often, as in many parts of the USA, they simply conferred the right to take water on the first person to do so, creating a hierarchy of property rights that became ever more difficult to monitor, record and administer. In more arid parts, the situation was increasingly reached where all the water in a region had been allocated (whether formally or informally), leading to the definition of some river basins as 'closed' (in other words, closed systems with no outflow to the sea or other water bodies, and where additional water needs could not be met without reallocating water from other users or by improving water use efficiency – activities which require more complex institutions with mechanisms for negotiation and conflict resolution).

Where single-purpose infrastructure was developed to serve, for instance, the farmers of a particular region, as in many parts of the Indian subcontinent, it was sufficient to establish an agricultural administration that could control irrigation water use along with other aspects of the schemes. But as more schemes were built, and other water uses and users emerged, it became increasingly difficult to enable continued use without coordination and engagement among different users.

Even where the needs of individual users could be sustained, this was often at the cost of the natural environment – and of communities who depended on the environment for their livelihoods, whether through fisheries or other products. As a result, the administrations formerly responsible for development and operation of 'hard' water resource infrastructure had to pay greater attention to the 'soft' management and protection of the resource.

It is pressures such as these that have led to the emergence of the concept of integrated water resources management, which in many cases reflected good practice rather than any startling new innovations. While this concept has its roots in these good practices and in the analytical work of the Harvard Water Program and others in the 1960s (see Chapter 14), formally the concept emerged as part of a package of approaches designed to achieve sustainable development that was adopted by the 1992 United Nations Conference on Environment and Development (UNCED), following the

publication of the Brundtland Commission's report, 'Our Common Future' (World Commission on Environment and Development, 1987).

The integrated approach to water resources management arose, in part, to help address challenges that traditional approaches to management could not cope with. However, to address these challenges, it was necessary to be clear about the criteria that would guide such water resource management efforts.

Individual user sectors often had clear criteria – drinking water supplies must be safe and water supplies for large industries (and for major transport routes) must be reliable. But where the resource is managed to the benefit of a number of different sectors, which criteria should apply? Which should get priority?

The response has been to recognize that there are multiple criteria that need to be used to guide the management of water for different uses. Within this, an important advance was the recognition that the maintenance of the water environment could be considered as a use, in itself, particularly where specific economic (e.g. livelihood) and social (e.g. recreation) services were derived from it.

This recognition did not come about in isolation. Driven by environmentalist concerns from the 1970s onwards (as evidenced by the World Conservation Union's (IUCN) 1980 'World Conservation Strategy' and the Brundtland Commission's 1987 'Our Common Future'), systematic attempts had already begun to address the need to understand development within an environmental framework, an approach that was finally formalized internationally by the 1992 Earth Summit (UNCED) in Rio de Janeiro. There, the general approach to what became known as 'sustainable development' encapsulated the same balancing act among environmental, social and economic priorities that is embodied in the concept of IWRM.

The IWRM approach was initially most comprehensively articulated in the chapter on freshwater resources in Agenda 21 of the Earth

Summit (see Box 1.1), which was informed by what became known as the Dublin Principles – a set of four basic principles produced at a Summit preparatory meeting in that city in 1992. IWRM can thus usefully be considered as the water element of the broader sustainable development approach.

Box 1.1   Agenda 21 provision for the application of integrated approaches to the development, management and use of water resources

The widespread scarcity, gradual destruction and aggravated pollution of freshwater resources in many world regions, along with the progressive encroachment of incompatible activities, demand integrated water resources planning and management. Such integration must cover all types of interrelated freshwater bodies, including both surface water and groundwater, and duly consider water quantity and quality aspects. The multisectoral nature of water resources development in the context of socioeconomic development must be recognized, as well as the multi-interest utilization of water resources for water supply and sanitation, agriculture, industry, urban development, hydropower generation, inland fisheries, transportation, recreation, low and flatlands management and other activities. Rational water utilization schemes for the development of surface and underwater supply sources and other potential sources have to be supported by concurrent waste conservation and wastage minimization measures.

Source: *Chapter 18, Agenda 21*

Agenda 21 explicitly promotes the use of the resource base in ways that best support social equity, economic development and environmental sustainability objectives. The IWRM approach reflects this concern in that it seeks to achieve an optimum balance among the 'three Es': efficiency, equity and environment. IWRM provides a way of operationalizing this part of Agenda 21, offering a problem-solving approach to address key water-related development challenges in ways that balance:

- economic efficiency – to make scarce water resources go as far as possible and to allocate water strategically to different economic sectors and uses;
- social equity – to ensure equitable access to water, and to the benefits from water use, between women and men, rich people and poor, across different social and economic groups both within and across countries, which involves issues of entitlement, access and control;
- environmental sustainability – to protect the water resources base and related aquatic ecosystems, and more broadly to help address global environmental issues such as climate change mitigation and adaptation, sustainable energy and sustainable food security.

To achieve this balance, it is useful to view the IWRM approach as the operationalization of what are often termed 'IWRM principles'. Our view is that these principles can be expressed very simply as the recognition that water is a public good with both social and economic values, and that good water resources management requires both a broad holistic perspective and the appropriate involvement of users at different levels.[1]

The meaning of integration

Although the Earth Summit emphasized the importance of getting water managers to take a more holistic approach to the resource and of bringing actors from different sectors into water decision-making processes, there was more to integration than encouraging users to work together. Once it was recognized that water needed to be managed as a contribution to broad economic and social development, it became clear that its planning and management had to reflect broader national priorities.

Critically, at a physical level, there was also recognition that the hydrological cycle is a unitary one and that the apparently separate bodies of water flowing in rivers and underground, falling to the earth as rain, accumulating in lakes and aquifers, and being evaporated from the earth's surface, are all interconnected. Thus, the way that land is managed has an impact on water resources and vice versa. Further, since the ability of the water resource to absorb the wastes that are dumped into it while continuing to sustain the ecosystems it supports depends to a large extent on how much is available; the quantity of water could not be managed in isolation from the quality of water.

But the meaning of the term integration was recognized as going significantly beyond integration within natural systems. It meant bridging such natural systems with the human systems that determine the demand side of the equation and development priorities. It meant bridging the water sector and other sectors of the economy. And, perhaps most importantly, it meant 'vertical' bridging across spatial scales and levels of decision-making – from local, provincial and national to water basin and transnational, with actions at one level seeking to reinforce and complement action at other levels.

The concept and challenges of integration are not limited to water. And integration is not an end in itself. While many sectors and areas of activity are related to others, it is seldom possible to manage them *all* as a single unit. Many different specialist institutions and organizations have been designed to cope with the demands of coordinating specialization. Ensuring effective coordination between specialized activities and institutions is a core element of the art of management, whether in public administration or large businesses.

Thus, the concept of integration does not entail trying to connect and manage everything together with everything else – a situation which would rapidly become unworkable. Nor does an integrated approach imply that sectoral decision-making needs to be abandoned entirely; on the contrary, achieving results usually requires some degree of targeting and focus. Trying to establish formal management relations between too many variables risks getting mired in complexity at the expense of effectiveness.

In this context, the key words in IWRM remain 'water resources management'. 'Integration' is simply the shorthand chosen to describe the kind of management that the approach entails – an approach that could just as accurately be described as 'holistic' or 'systemic'.

Complexity and diversity: blueprints, best practices and the key features of IWRM

With this background, it should be clear that there can be no uniform, blueprint approach to water resources management. Development priorities and social and economic challenges differ from country to country, as do water resource endowments and levels of infrastructure, the opportunities to make changes, and the institutions of law, custom and practice. Even within countries there are often significant regional differences that shape water resource challenges and possible solutions.

Thus, while the overall approach may be common, in its application it can take many forms, and each community, basin and country needs to determine the approach most suited to its own particular context. The best path will look very different in different countries, and in different regions, basins and communities within a country. As an example, in their response to the Johannesburg Summit's call for all countries to prepare IWRM and Water Efficiency Plans, some countries have prepared formal IWRM plans and policies, others decided that their existing water resources management policies and practices were adequate, albeit not codified in a formal plan (United Nations Economic and Social Council, 2008).

But while there can be no blueprint, experience has shown that effective strategies for better water resources management consistently include some common features. Good practices almost certainly involve, to one extent or another, the following elements:[2]

1 sound investments in infrastructure – to store, abstract, convey, control, conserve and protect surface and ground water;
2 a strong enabling environment – setting goals for water use, protection and conservation; improving the legislative framework; enhancing financing and incentive structures; and allocating financial resources to meet water needs;
3 clear, robust and comprehensive institutional roles – laying out institutional forms and functions, building institutional capacity, developing human resources, establishing transparent processes for decision-making and for informed stakeholder participation;
4 effective use of available management and technical instruments – for such purposes as water resources assessment, water resource management planning, demand management and social change, conflict resolution, allocation and water use limits, using value and prices for efficiency and equity, information management and exchange.

It is this basic package whose effectiveness is illustrated, in one way or another, by the case studies in this book.

Our conceptual framework

The book takes a case study approach. Collectively, the cases illustrate the many different contexts in which IWRM principles have been applied, the different approaches that have emerged, and the results that have been achieved. Because the objective of the IWRM approach is not water management as such but human development, the examples focus as much on the development challenges as on the water challenges themselves. The cases not only deal with cross-sectoral development issues but also address major 'single sector' challenges, such as enhancing food security through irrigated agriculture.

Importantly, none of the cases described in this book was conceived as 'an IWRM project' and none had an integrated approach as its principal objective; rather, the integrated approach became necessary to address the specific development problem in the case at hand. Indeed, we have explicitly included several cases that had their origins well before the concept of IWRM was formally adopted by the 1992 UNCED in order to reinforce the fact that the emergence of the IWRM concept reflected prevailing good practice rather than a radical new direction.

The cases

The cases demonstrate IWRM at a variety of scales, from small initiatives such as the village-level Sukhomajri case in India, to large transboundary ones such as the Mekong Commission, which covers four countries and seeks to promote collaborative management of a river that nurtures a huge portion of southeast Asia. The examples have also been deliberately selected to illustrate how the problems and possible responses change at different levels of economic and social development. And the cases highlight the importance of addressing all three aspects of the IWRM approach – social, environmental and economic.

Critically, and almost by definition, an IWRM approach can only work if it does not focus exclusively on water. So massive water resource management initiatives in China, such as the development of the Yangtze, which includes the construction of the controversial Three Gorges Dam, need to be seen in the context of the social and economic challenges faced by society.

The cases from South Africa and Chile show societies where major social transformations have opened new and unexpected opportunities to allow water management to support the achievement of broader social and economic goals. In both cases, political upheavals created the opportunity to put in place new water management frameworks that are better able to cope with the challenges which the countries face.

In many of the cases it is social transformation and development that are driving the need to manage water better. Thus, in Mali, the challenge faced is to ensure that water management supports the ambitious goals of national development. In Bangladesh, the challenge was to turn an overabundance of water from a development constraint into a development opportunity. This case also shows that development and environmental goals need not be in opposition – it is possible to improve the incomes of poor rural communities while sustaining delicate coastal wetlands.

Several of the cases from industrialized countries demonstrate the high cost and difficulty of undoing previous damage to the environment. Rich countries such as Japan have had to find ways to maintain their economies even as they begin to enforce environmental protection measures to address the impact of decades of industrialization on inland lakes. The Japanese case also shows how, as societies grow and develop, their priorities often change and approaches that met their needs in the past may not be acceptable in the future. The debate amongst the communities along the Snake river in northwest USA, where proposals have been made to remove dams and other water infrastructure and return the river to nature, is a further illustration of this process. All these examples highlight that an IWRM approach is not a recipe or a one-off formula. As a community or country changes, so too will its water management challenges and responses.

It is critical to recognize this natural evolution and to ensure that the institutions, the 'hard' formal organizations and water management structures, the 'soft' institutions of law, custom and practice are designed so that they facilitate change in their societies rather than create impediments to it, as has too often happened in the past. These questions are explored in the case on the establishment of a River Basin Council for the Lerma–Chapala River Basin

in Mexico, where the different interests of 15 million people, their local governments, industries and agriculture had to be reconciled.

If it is already a challenge to build institutions at the level of one national river basin to help the parties who share its water to work together constructively, how much bigger is the challenge at the level of rivers that are shared between many different countries? The case study from South Africa, which shares many of its rivers with its neighbours, and that concerning the Mekong in Asia, look at the progress, often painfully slow, that has been made to move from a situation where there has been conflict to one where historic antagonists are now working together for mutual benefit, with water often a catalyst for cooperation.

The elements of our conceptual framework

Our conceptual framework includes five elements: the varying levels and scales of the problem and response, the development context, the changes in policies and practices embodied in the response, the development outcomes of these changes, and the learning that ensues. We discuss each of these five elements below.

Scale

We have chosen to use scale as an organizing principle for the case studies in this book. Needs and opportunities arise at many different levels or scales, from isolated rural communities for whom better water management can provide electricity as well as irrigation opportunities, to whole countries whose social and economic prospects can be transformed and where cooperation between nations over water management can unlock win–win opportunities that reduce tension, enhance security and promote broad economic well-being.

Development and natural resource management processes also occur at different scales – from local, provincial and national to water-basin and transnational scales. Stresses manifest themselves in different ways at different levels and may be addressed bottom-up or top-down depending on the nature of the problem and the development context. But as demonstrated by many of the selected cases, action at one scale must reinforce and complement action at other scales. Action at the national scale, for example, can and should provide the strategic framework for actions at local levels.

Given the central importance of scale, the cases in this book have been ordered on the basis of their scale, from local to transnational. This is also intended to make the book easier to use. Those interested in community-based interventions or local government will probably find the cases at the local scale more applicable to their work, while policy-makers and international development organizations may find the regional, national and international cases more relevant.

Development context

As emphasized earlier, development priorities and water resource challenges and practices differ considerably from country to country, and within countries. The particular changes needed to achieve an optimum balance among economic efficiency, social equity and environmental sustainability depend fundamentally upon these contextual factors. Thus, the actions taken to improve water management and the sequence of those actions will vary dramatically from one place to another and will change over time.

Given the importance of context, cases have been selected to depict a range of different social and economic challenges. The cases also take place in countries with differing levels of water resource endowments and levels of infrastructure, and with traditions of governance that range from strong centralism to extreme federalism. Many cases describe geographic locations within countries, whose distinct regional characteristics shaped the water management challenge and outcomes.

Policy and practice change

All efforts to improve water management involve changes in policies and practices. To minimize external costs, achieve economies of scale and scope, reduce cross-sectoral competition and improve developmental outcomes, more integrative and people-centred approaches are becoming imperative. And meeting future development and environmental sustainability challenges, especially in the face of long-term climate change, will require further changes in the way in which water resources and water services are managed.

Experience to date, however, suggests that deliberate change (as opposed to passive processes of evolution) is inherently difficult to achieve. Many constraints stand in the way, and formidable forces frequently gather to maintain the status quo even when there is increasing pressure from a changing natural and economic environment. A prescription for change cannot simply be transferred or imposed by fiat. Change will only be successful if it:

- arises in response to a recognized problem or crisis, such as the need to accelerate provision of adequate and cost-effective water supply and sanitation services, or to reduce the frequency and impacts of floods and droughts;
- is socially, economically and technically appropriate to the particular context; and
- is grounded in existing institutions – tailored to current capacities and stage of development, and with attention to potential losers in the change process. Trade-offs must be taken into account in any process of change. While existing vested interests need not always dominate, it is usually helpful if all stakeholders perceive that the benefits of change outweigh the potential short-term losses.

The cases in the book describe and discuss the changes in policies and practices (in laws, organizational structures, and so on) that were an inherent part of the approach – for example, what was undertaken and how, the instruments used, the way in which efficiency, equity and sustainability considerations were addressed.

Development outcomes

Through the cases in the book, we show that applied appropriately an integrated approach does produce tangible and positive development outcomes. However, because of the many factors involved assessing outcomes and impacts in the context of water and development is complicated. In addition, as the cases have shown, returns from investments in water frequently come back not as income to the originating entity, such as a water utility, but as wide-ranging and often long-term benefits spread across different segments of the economy, making measurement more difficult. Just as it is more difficult to evaluate the impact of a generally healthy lifestyle than a specific treatment on an individual's health, so it is more difficult to evaluate the impact of a broad approach than a specific project on a country's development performance.

Because of these difficulties, previous IWRM case studies have tended to focus on processes – changes in policies, laws, organizational structures – with little attention to ultimate outcome and impact. While many studies have shown that good water management has positive benefit/cost ratios, and rigorous analyses across many disciplines have demonstrated the benefits of multi-scale, multi-sectoral and multi-objective decision-making that is relatively more long-term in its planning horizon, there is little literature to which one can direct decision-makers who are interested in knowing what returns they can expect if they adopt the kind of integrated approaches to water and development outlined in this book.

Despite these challenges, we have tried to ensure that the cases in this book demonstrate some tangible impact on economic growth, social equity and/or environmental sustainability – going beyond the traditional focus on process-related changes to show concrete impacts.

Learning

The aim of this book is to learn about IWRM, and about water and development more generally, from those who have already 'done it'. One of the criteria

for selection was the extent to which the case provides useful experience and lessons for other countries, communities and situations. We have tried to highlight the lessons in each case that would be useful in a wider context.

It is widely recognized that practical experience is the best teacher; a great deal of learning is about learning from our mistakes. As important, we believe, is learning from our successes and one feature of the IWRM approach described in this book is that it promotes processes of collaborative work through which people from different sectors can together seek solutions to their water challenges – and thus 'learn by doing'. We hope this book will also offer such people – both policy-makers and practitioners – an opportunity to learn from the mistakes of others rather than repeating them endlessly themselves.

Extracting the lessons

There are great challenges in deriving overall conclusions about what works and what doesn't from the diverse cases that we have presented, emerging as they do from varied socioeconomic and environmental conditions with outcomes that are variable and measurements that are not necessarily consistent; in other words, from cases that have no controls to help isolate the factors that contributed to success or failure. Nonetheless, we believe there is enough information to provide a valuable start. We have therefore distilled from the case studies a set of key messages, which are drawn together in the concluding chapter but can be summarized briefly as follows:

The objectives and outcomes of good water management

The appropriate balance among social, economic and environmental objectives is largely determined by a country's values and national development priorities – these priorities change over time, requiring water management structures that can adapt. Better water management not only impacts the water sector, but can also further economic growth, poverty reduction and environmental sustainability. However, 'optimizing' economic growth, social equity and environmental sustainability implies that there will be compromises and tradeoffs. While this is most often true, it does not follow that there is always a contradiction between the protection of the environment and promotion of economic and social development. As many of the cases in this book demonstrate, win–win situations are possible.

What constitutes good governance and water resources management

Managing water effectively requires the sustained collective effort and engagement of people in all sectors of society if it is to be successful in achieving the society's goals. It needs robust, competent and trusted institutions as well as economically, socially and environmentally sound investments in infrastructure. Pragmatic, sensibly sequenced institutional approaches, which respond to contextual realities, have the greatest chance of working in practice; but policy reforms and their implementation will only succeed if underpinned by a sound technical foundation.

How the management of water differs at different scales

Water resources planning and management must be linked to a country's overall sustainable development strategy and public administration framework. Better management of water at local level often needs the support of a sound policy framework at regional and national levels. In large river basins, effective governance from local to basin levels is a major challenge, requiring functions to be placed at appropriate levels; but while a river basin perspective is vital, it must often be supplemented by overarching national policies if water management is to be effective. Transnational governance is a special case requiring specific approaches.

The nature of the IWRM approach itself

IWRM is an approach rather than a method or a prescription, and there is no 'magic bullet' for all situations. Successful IWRM efforts adopt an integrated approach in order to address specific development problems; they never have an integrated approach as their principal objective. And finally, the process of water management does not have an end point and will continually have to respond to new challenges and opportunities.

In learning lessons from the case studies – each in its own right, but against a backdrop of common patterns and themes – we are helped by the natural cycle of water. Through the repeated cycles of the seasons, we have to respond to the challenges of water which are presented, slightly differently, each year. So water can be a natural teacher for those who are willing to learn.

Notes

1. This summary draws in part on Global Water Partnership, Conditions for Accreditation for Regional and Country Water Partnerships, 2007. While the first two principles are based on those in the Dublin Statement on Water and Sustainable Development, 1992, the third one is formulated to reflect the wording of Chapter 18 of Agenda 21.
2. These elements of good practices have been drawn in part from Global Water Partnership Technical Advisory Committee, 2000; Global Water Partnership, 2003; and Lenton, Wright and Lewis, 2005.

References

Brown, C. and Lall, U. (2006) 'Water and economic development: The role of variability and a framework for resilience', *Natural Resources Forum* vol. 30, pp306–317

Global Water Partnership (2003) *Integrated Water Resources Management Toolbox, Version 2*, Stockholm, Global Water Partner Secretariat

Global Water Partnership Technical Advisory Committee (2000) *Integrated Water Resources Management. TAC Background Papers* No. 4: 1–71, Stockholm, Global Water Partnership

Lenton, R., Wright, A.M. and Lewis, K. (2005) *Health, Dignity and Development: What Will it Take?* Report of the UN Millennium Project Task Force on Water and Sanitation, London, Sterling, VA, Earthscan

Matas, C.R. (2001) 'The Problems of Implementing New Public Management in Latin American Administrations: State Model and Institutional Culture', *Revista del CLAD Reforma y Democracia*, vol. 21 (accessed at) http://www.clad.org.ve/portal/publicaciones-del-clad/revista-clad-reforma-democracia/articulos/021-octubre-2001/0041100

Moorehead, A. (1962) *The Blue Nile*, London, Hamish Hamilton

Smakhtin, V.U., Revenga, C., Döll, P. (2004) Taking into account environmental water requirements in global-scale water resources assessments. Research Report of the CGIAR Comprehensive Assessment of Water Management in Agriculture. No. 2, International Water Management Institute, Colombo, Sri Lanka, 24 pp

Smith, A. (1776) *The Wealth of Nations*, 1986 edition, London, Penguin

Standard and Poors (2005) 'US Economic Update: Impact From Katrina Big, But How Big?' Available from: *www2.standardandpoors.com/spf/pdf/fixedincome/09–01–05_USEconUpdate_ImpactFromKatrina.pdf* (accessed 1 September 2008)

UNCED (United Nations Conference on Environment and Development) (1992) *Agenda 21, Report of the United Nations Conference on Environment and Development.* Available from: www.un.org/esa/sustdev/documents/agenda21/index.htm (accessed 4 June 2008)

United Nations Economic and Social Council (2008) *Review of progress in implementing the decision of the thirteenth session of the Commission on Sustainable Development on water and sanitation. Report of the Secretary-General*, Commission on Sustainable Development, 16th session, 5–16 May 2008, item 4 of the provisional agenda. Document No. E/CN.17/2008/11

Waterston, A. (1965) *Development Planning: Lessons of Experience*, Baltimore, Johns Hopkins University Press

World Bank (2004) *Towards a Water-Secure Kenya*, Water Resources Sector Memorandum, vol. 1 of 1, Washington, DC, World Bank

World Commission on Environment and Development (1987) *Our Common Future* (Brundtland Report), New York, Oxford University Press

Part One – Local Level

Water management is often a local affair, whether it involves a few farmers working together to bring water from a stream to their fields or a village's citizens seeking a safe and reliable source for their public water and a way to dispose of their wastes.

The practices that are established at this level can build a foundation for wider cooperation in the future. However, they can also sow the seeds of future conflict or, at the very least, environmental damage.

The examples selected here illustrate how local cooperation and management can grow into something bigger. Three of them focus on agriculture, which is by far the largest water user in most countries. The cases of Sukhomajri in India and the Office du Niger in Mali show how farming livelihoods can be improved through better water management. On the other hand, both the Sukhomajri case and that of Aalborg in Demark show that the way farmers manage their water can have significant impacts on water users in other sectors. In both cases, urban residents could not obtain adequate water supplies without engaging with the production methods of the farmers in their areas.

But the impacts of water management go well beyond agriculture and drinking water. The case from Bangladesh shows how the livelihoods of fishermen improved as a result of some strategic management innovations. There, livelihood benefits were accompanied by clear environmental gains. Another such 'win–win' occurred in the Australian Angas Bremer case. But such solutions cannot always be found; sometimes there have to be difficult tradeoffs between different objectives, as in the case of the Snake River in the USA where water use for power generation and transport conflict with environmental conservation efforts.

These local cases also illustrate the point that water challenges are often best resolved through action in other sectors, as occurred in Mali where improved water-use efficiency was achieved by addressing the institutional arrangements for agriculture in the Office du Niger irrigation scheme. And, as the Angas Bremer case shows, local problems can sometimes only be solved with support from the wider community.

Finally, these local cases show clearly the links between different levels of action. Several of the local-level cases, in particular Sukhomajri and Office du Niger, have had significant national implications. And most highlight the need for actions at the local level to be supported by actions at other levels, and especially by a supportive policy framework at regional and national levels.

2

A Watershed in Watershed Management: The Sukhomajri Experience

Roberto Lenton and Christie Walkuski

Watershed management has been defined as 'the integrated use of land, vegetation and water in a geographically discrete drainage area for the benefit of its residents, with the objective of protecting or conserving the hydrologic services which the watershed provides and of reducing or avoiding negative downstream or groundwater impacts' (World Bank, 2007). It has played a prominent role in rural development efforts in many countries in the last several decades, helping to increase rural incomes, augment usable water resources, improve productivity and mitigate droughts. Watershed management programmes have been driven both by the desire to protect downstream water facilities as well as to support livelihood generation and environmental conservation in the watershed itself.

A high degree of interconnectivity is a crucial feature of watershed management. Upstream/downstream and rural/urban linkages in particular are often critical, particularly in areas of rapid urbanization and where land and water resources are increasingly stressed. Poverty, land degradation and erosion in upstream areas can lead to downstream floods, poor water quality and sedimentation. Water is the central link and the starting point. Addressing the needs of the communities who live in the watershed and ensuring shared benefits is therefore the key to successful watershed management. A focus on the watershed level helps to develop solutions that will preserve and protect natural resources, improve soil quality and water supply for agriculture and human needs, and increase economic and social opportunities.

While watershed management arose independently of the concept of integrated water resources management, its focus on the coordinated development and management of water, land and related resources in a watershed – and its ultimate aim to conserve natural resources as well as improve livelihoods and reduce poverty in an equitable way – mirrors both the means and the ends of an IWRM approach. The extensive experience with watershed management programmes over the last several decades thus provides an outstanding opportunity to examine firsthand the results of an integrated approach that has been applied widely in different contexts.

The Sukhomajri programme in northwest India is an example of watershed management that has yielded strong and sustained development impacts for close to three decades. The experience has been extensively documented and analysed, both in its initial stages (Franda, 1981; Seckler, 1986) and over time (Agarwal and Narain, 1999; Kerr, 2002; Narain and Agarwal, 2002; Khurana, 2005). In addition, the Centre for Science and the Environment (CSE) in New Delhi has kept a watching brief on the experience since 1994, as reported in CSE (1994), CSE (1998), CSE (2002) and CSE (2007).

The programme's 30-year history is long enough to permit a serious analysis of sustainability considerations in the light of the 'second-generation' challenges that typically arise over the long haul.

The development context

As with many early watershed management programmes, the Sukhomajri programme began as a result of the connections between a major urban centre, Chandigarh, and a rural upstream village, Sukhomajri. Floods, poor water quality and sedimentation downstream in Chandigarh were found to be linked to poverty, land degradation and erosion in Sukhomajri, and it was these linkages that gave rise to the programme in the 1970s.

Chandigarh, which is the capital of two states – Punjab and Haryana – in northwest India, is a modern city with a high per capita income and standard of living. Planned by the famous architect and urban planner Le Corbusier during the early years of Indian independence, Chandigarh was viewed as a model 'beautiful city' and enjoyed a greater level of investment and development than its neighbouring villages. In 1958 the city created its own Sukhna Lake by damming the local Sukhna Choe, to serve as water supply and as a recreation area.

The village of Sukhomajri is located approximately 15 kilometres northeast of Chandigarh, at the headwaters of Sukhna Lake, on the edge of the Shivalik mountain range in the state of Haryana (see Figure 2.1). The climate in the region is semi-arid with low rainfall, most of which occurs in the monsoon months of June through September (Franda, 1981; Agarwal and Narain, 1997). Like many villages and towns in the sub-Himalayan foothills, Sukhomajri inherited the problems caused by over a century of heavy logging in the area and the overgrazing of cattle, sheep and goats in open forest lands. Before the start of the Sukhomajri programme, the people were very poor and survived primarily by raising rainfed crops and keeping goats, which were able to forage in the denuded hills. Water supply came from a rainfed pond and a nearby spring, but there was no irrigation water for the fields (Seckler, 1986).

Sukhomajri was caught in a cycle of poverty and environmental degradation. Population growth and limited livelihood opportunities contributed to overgrazing and deforestation and thus more erosion. Where once there had been lush, forested slopes, gullies and clay banks became increasingly prominent. Rugged terrain and sandy soils exacerbated runoff and erosion. At one end of the village several acres of good cropland disappeared into a gorge. Farmable land was decreasing. Seasonal rains continued to wash sediment down towards the plain below the hills and into Sukhna Lake in Chandigarh (Agarwal and Narain, 1999; Kumar, 2003).

At the start of the Sukhomajri programme the village consisted of 59 families (approximately 450 people), all of the same Gujar caste. Most lived in mud and thatch houses and owned less than one hectare of land. The village land comprised some 100 hectares, half of which was owned communally. The Forest Department owned much of the surrounding area, including about 400 hectares of denuded forestland that was used by the community as grazing land. Crop production was insufficient and most families kept goats to supplement their incomes. Both food and fodder were imported from other villages (Franda, 1981: 6; Narain and Agarwal, 2002: 9).

During the 1970s, it became increasingly clear that the fates of Sukhomajri and Chandigarh were intertwined, principally because the sediment washed down by seasonal rains in the Sukhomajri watershed rushed down into Chandigarh's Sukhna Lake. By the early 1970s the lake had become so choked with sediment that it had lost nearly 70 per cent of its storage capacity and boat owners couldn't even get their crafts out of the harbour. The city was spending US$200,000 per year on dredging operations that proved useless (Seckler, 1986: 1016; Kerr, 2002). City officials turned to the Central Soil and Water Conservation Research and Training Institute (CSWCRTI), based in nearby Dehra Dun, for help, and the Sukhomajri programme was born.

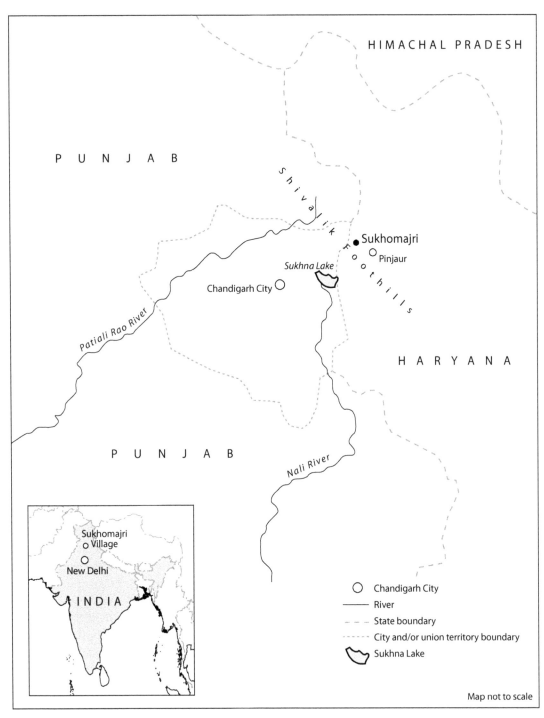

Data sources: CIESIN

Figure 2.1   Sukhna Lake, Sukhomajri village and the Shivalik foothills

The Sukhomajri approach: the initial stages

Initial efforts by CSWCRTI and the state Forest Department to manage the watershed focused on erecting erosion control structures outside of Chandigarh. While these measures helped reduce some of the problems, it was soon learned that the hills in and around Sukhomajri, particularly one severely eroded ravine, were largely responsible for the torrent of sediment coming down. Attention turned to the small village for a solution. But simply employing erosion control mechanisms was not going to be enough.

Not surprisingly, CSWCRTI was met with resistance when it started its work in Sukhomajri. While CSWCRTI's aim was to improve soil and water conservation to protect resources and interests downstream, the people in Sukhomajri saw no reason to be concerned about Chandigarh's dying lake. Their principal concern was their economic survival, which depended on the continued use of the watershed to graze their animals.

As Seckler notes (1986), the breakthrough – for Sukhomajri and indeed for Chandigarh – came when CSWCRTI built a small earthen dam to control runoff, and the villagers suddenly had a reliable source of water relatively close at hand. For a very dry village, water was wealth. When water collected in the reservoir and didn't seep into the ground or evaporate away, the villagers and the institute saw a joint opportunity and began to cooperate. Their successful mutual engagement was largely predicated on three factors: community participation in decision-making and management incentives for villagers to graze their animals outside of the watershed (in essence, payment for conservation services of the watershed) and a system of water allocation that would benefit all villagers equally.

P.R. Mishra, the strong and charismatic director of CSWCRTI's Chandigarh office at the time, understood that the project needed to be owned by the villagers and that any benefits needed to be shared. Here were the beginnings of a truly participatory approach: government agency staff and village farmers and herdsmen started working together to implement practices aimed at protecting water, land and forest resources. Community participation and an equitable distribution of benefits were recognized as necessary components (Franda, 1981). According to S.P. Mittal, a scientist who started the project with P.R. Mishra, 'In many ways the Sukhomajri experiment was the first instance of challenging the top-down model of development' (CSE, 2007).

The initial check dam that was built was about 10 metres high. Using the stored water, villagers found that during a drought they were able to save wheat crops growing close to the dam. Three more check dams were constructed, all piping water via gravity to the fields below. With support from a Ford Foundation grant, CSWCRTI provided the initial funding for the check dams and underground pipes and for the erosion control methods built throughout the watershed. Farmers were also provided with fertilizers and high-yielding seed varieties and were encouraged to level their land to improve water distribution on their fields. They were more than willing to do so and to bear the cost of these measures themselves. Construction of soil conservation mechanisms within the watershed continued with staggered contour trenches, grade stabilizers and gully plugs, all constructed by CSWCRTI. With the water came increased crop yields and stabilized agricultural production, as well as the incentive to expand the area of land under cultivation.

In exchange for the water from the check dams, villagers were willing to graze their animals in lands outside of the watershed. But there was a problem: although the water was able to reach many of the village fields through the piped conveyance system, it did not reach all of them. Tension mounted. The irrigated crops succeeded and the non-irrigated crops failed. Some villagers continued to graze their animals in the watershed. A solution was needed.

The villagers organized and formed a water users' association, later to become the Hill Resource Management Society (HRMS). The HRMS was given authority to manage the dams and distribute the water. At first, this was done via a coupon system, which villagers were able to sell if they wanted. Water rights were granted regardless of whether villagers owned land or not. Thus, everyone had a share of water and everyone had an incentive to keep up their end of the bargain.

Just as water gave the villagers a stake in protecting the watershed, the right to cut grass growing on Forest Department land on a per unit basis gave villagers a stake in protecting the forest. The villagers used mungri grass as livestock fodder and bhabbar grass could be sold to the local paper company, which used it for pulp. Previously, the Forest Department had auctioned grass-cutting rights to a contractor who then would sell the grass to the villagers at a high unit price. HRMS negotiated leases for mungri and bhabbar grasses and was able to sell to villagers at a much lower unit rate. These arrangements thus provided an income stream to HRMS, based on the sale of both fodder to villagers and bhabbar to the paper industry, which was then cycled back into the community (Narain and Agarwal, 2002; Kumar, 2003).

Economic efficiency, environmental sustainability and social equity

Sukhomajri's holistic, community-based approach to developing and managing water, land, forest and livestock in the watershed aimed at achieving a good balance of economic efficiency, social equity and environmental sustainability.

The water allocation scheme HRMS devised was based on the traditional principle of *warabandi*, or 'water turn', widely used in irrigation schemes across northwestern India and Pakistan, under which farmers are allocated water based on irrigable land area and crop type. This water allocation system has two key principles: farmers receive considerably less water than they could potentially use, creating a scarcity of supply and thus encouraging highly efficient water use on the farm; and water is equitably distributed to all land holders, ensuring that the maximum number of such land holders benefit from the total supply.

The Sukhomajri scheme took this latter principle one step further by allocating water not only to land holders but also to the landless; indeed, it was decided that each household would get an equal share of water, regardless of whether they owned land or not. The water shares were also tradable so that those who didn't need the water could sell to those who did, or water could be traded for sharecropping arrangements. HRMS administered all transactions and distributed all funds. To encourage efficient use, fees were charged based on the amount of time water was supplied to the farmer. The measures taken by the villagers to improve cropland and agricultural production, such as land levelling and installing their own erosion control works, also no doubt improved water efficiency. Soil improvement efforts also enabled crops to flourish with less water.

Efforts to distribute water efficiently and equitably also had important environmental protection benefits. Incentives for villagers to graze their animals outside of the watershed helped address poor watershed conditions and the pervasive problems of severe runoff, erosion, sedimentation and deforestation. Upstream erosion prevention led to the protection of water infrastructure downstream, thereby saving maintenance, repair and replacement costs.

To achieve this balance of economic efficiency, social equity and environmental sustainability, the Sukhomajri programme included a mix of technical, institutional and financial measures:

• Technical measures: the check dam, irrigation, high yielding varieties, fertilizer, land levelling, watershed protection, erosion control works, livestock control, grass planting. Different parties (outside groups, the community, and individual farmers) took the lead in implementing each of these measures.

- Institutional measures: the development of the water users' association, the system for sharing water and other resources.
- Financial measures: the arrangement which allowed the HRMS to earn income from forestry leases and for dam and pipeline maintenance and forest protection, the provision for check dam construction costs to be covered by outside groups. (Farm level costs, such as livestock control and land levelling, were paid by individual farmers.)

Importantly, although the approach focused principally on village and watershed level actions, these were supported and complemented by actions at district and national scales. At the district level, efforts initially involved cooperation between departments for forestry and water and soil conservation. National-level efforts included the extension of the approach to other areas; Sukhomajri was one of the first examples of community integrated watershed management in India and has become a model for watershed development elsewhere in the country. At the national level, lessons learned from Sukhomajri and other similar projects were institutionalized in the form of the National Watershed Development Programme for Rainfed Areas and the formulation of common guidelines for watershed management projects, which were used by the Ministry of Rural Development and the Ministry of Agriculture. In addition, India's Integrated Watershed Development Programme funded by the World Bank, which ran from 1990 to 2005 in the five states in the northwestern Shivaliks including Haryana, was inspired by the Sukhomajri model (CSE, 2007).

The Sukhomajri approach: its evolution over time

Over the three decades since the programme was initiated, Sukhomajri has undergone many changes. While the initial stages transformed life in the village

and led to significant greening of the watershed, two important later developments negatively affected some of these gains.

The first and perhaps most important of these later developments was that the shallow tubewells, which villagers started digging in the early 1980s, began over time to replace water from the check dam and its pipelines as the most important source of irrigation water for the village. By 2001 around 17 hectares were irrigated by tubewells; meanwhile, the area irrigated with check dam water fell from 33 hectares in 1985 to only 10 hectares in 2001, a figure that by 2007 had been further reduced to 5 hectares (see Figure 2.2) (CSE, 2007).

Indirectly, the check dams themselves were a major cause of this development, since they helped raise groundwater levels from about 120 metres below ground before the dams were built to about 40 metres in 1981–1982. The higher water tables made it possible for individual farmers to tap into a plentiful and relatively cheap source of water that was fully under their control, so they were no longer dependent on the water shares allocated through the programme. Thus they had fewer incentives for limiting watershed grazing and participating in communal activities. But in the absence of a system to regulate groundwater withdrawals, the water table level began to drop again as a result of overextraction – by 2006 it had gone down to 90 metres (see Table 2.1).

Table 2.1   *Changes in groundwater levels*

Year	Groundwater depth
1979	120m
1981	40m
2005	90m

The second major development to cause significant setbacks was the Forestry Department decision to begin taxing the sale of bhabbar and fodder harvested on its land – initially at 25 per cent and then in 1998 at 55 per cent. Consequently, HRMS's income from the sale of bhabbar and

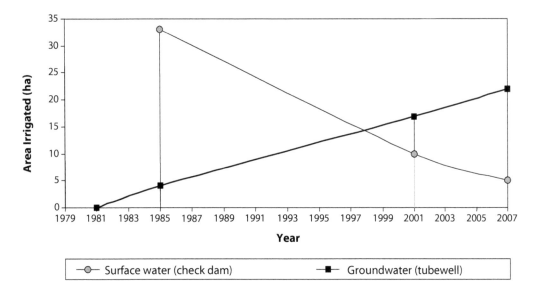

Source: CSE, 2007

Figure 2.2   Changes in area irrigated by different resources

fodder began a serious decline in net terms, starting in 1990. To make matters worse, the demand for bhabbar dropped precipitously as the paper industry began to use wood pulp instead, further reducing the Society's income. As a result, HRMS's annual income declined in the early years of this millennium to less than one-tenth of the level it was in 1986–1987, the first year HRMS earned revenues (CSE, 2007). Figure 2.3 shows the ups and downs in HRMS's financial fortunes over the last 25 years.

Due to the fall in HRMS revenues and the weakened incentives for tubewell irrigators to participate in watershed conservation, the dams silted up and the pipelines began to fall into a state of disrepair.

Nevertheless, while the above two relatively recent developments have taken their toll, Sukhomajri is still a prosperous village. And the outlook for the future depends greatly on the policy decisions that are taken going forward, both at the village level by individual farmers and the HRMS, and at higher levels by the Forest Department and the State of Haryana. The neighbouring village of

Bunga, for example, which initiated a similar watershed development programme after visiting Sukhomajri in the early 1980s, has been able to avoid Sukhomajri's recent problems, largely because its HRMS has more financial independence and is more assertive in controlling the use of common property resources (CSE, 2007).

Development outcomes

The Sukhomajri programme had important economic, environmental and social outcomes. In agricultural production terms, the dramatic increases in crop yields are by now famous: between 1977 and 1986, wheat production grew from 40.6 tonnes to 63.6 tonnes, and maize production grew from 40.9 tonnes to 54.3 tonnes (Narain and Agarwal, 2002: 11). Total cultivated land in the village also rose, from about 50 hectares in the mid-1970s at the project's start to 110 hectares in 1993 (Khurana, 2005). The boom in the staple crops wheat and maize improved agricultural stability,

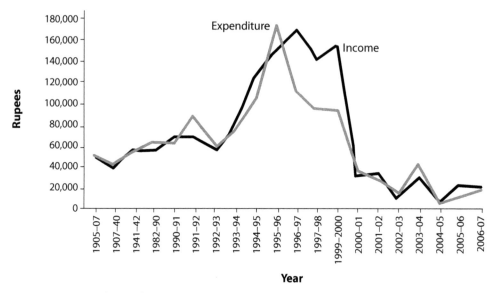

Note: Income figures after 2000 do not include balance carried forward
Source: Haryana Forest Department tax registers and personal communications

Source: CSE, 2007

Figure 2.3   HRMS income and expenditure, 1986–2006

and, as subsistence was assured, farmers began to diversify to more lucrative vegetable crops – such as peas, garlic, onion and potato – on a self-help basis, with no subsidies. Sukhomajri now exports rather than imports food.

Grass used for fodder grew 75-fold in the Sukhomajri programme's first decade, increasing opportunities for raising buffalo. Thus milk production grew from 334 litres in 1977 to 579 litres in 1986 (Narain and Agarwal, 2002). Despite the recent downturns, milk is still an important part of the village economy: 3,000–4,000 litres of milk worth 12 Rupees (US$0.30)[1] per litre are produced each day from a herd of nearly 560 buffaloes and cows (CSE, 2007).

Annual household income rose with crop, fodder and livestock production, climbing from around 10,000 Rupees (US$230) in 1979 to about 15,000 Rupees (US$340) in 1984, a 50 per cent increase in just five years (see Figure 2.4). According to CSE (2007), household incomes reached around 40,000 Rupees (US$910) in the 1990s and about 60,000

(US$1360) in the 2000s. Even after accounting for inflation, these are significant gains, and reflect the fact that, in a relatively short time period, villagers have had more and more opportunities to improve their livelihoods and their economic wealth. As emphasized by Narain and Agarwal (2002: 1), economic wealth was created and built upon natural wealth and capital: water, crops, animals and timber. Addressing ecological poverty was akin to addressing economic poverty. As one villager put it, 'Who could imagine that televisions, tractors and bicycles could be had for mere grass and water?'

In spite of recent developments, Sukhomajri still boasts more than double the per capita income of the state of Haryana, which is one of India's more prosperous rural states. Villagers now live in brick and cement houses and enjoy modern amenities such as cars and televisions (Narain and Agarwal, 2002: 12; CSE, 2007). Families are also accruing savings, and per capita consumption data show expenditures on clothes, health care and other non-food items totalling more than 20 per cent of overall

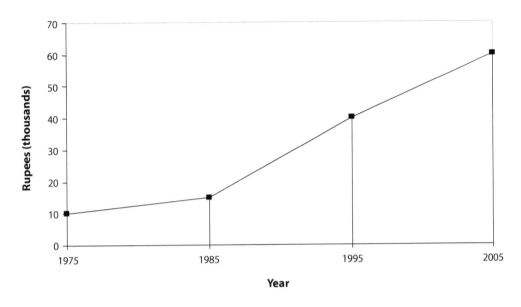

Source: CSE, 2007

Figure 2.4   Changes in household income

spending, pointing to a higher standard of living than existed previously (Khurana, 2005).

But the economic outcomes of the Sukhomajri programme are only part of the story. Environmental outcomes from Sukhomajri have also been significant. In particular, watershed protection, including the careful regulation of grazing, resulted in an increased tree density in the area from 13 per hectare to around 1,300 per hectare in the 16-year period extending to 1995. This not only provided another source of income but also reduced soil erosion, which in turn reduced siltation of downstream reservoirs. In addition, by 1990 farmers, using their own labour, had constructed some 300 kilometres of erosion works.

From a social perspective, perhaps the most important outcome of the Sukhomajri programme has been its equity impact, especially regarding the income of landless people. In addition, outmigration largely ceased, at least in the early years, and the landless in the area benefited from an increase in the daily wage. Furthermore, the HRMS

has helped advance community development. Not only did the Society maintain the dams and other irrigation and anti-erosion infrastructure, it also managed irrigation and fish-farming enterprises in the reservoirs, and implemented a number of development activities within the village. Schools were built and roads were improved. While this institution has gone through difficult times of late, it has clearly played a major role in the social development of Sukhomajri over the last several decades.

Importantly, the development outcomes of the Sukhomajri programme have occurred at more than one level, since they have extended significantly beyond the catchment area. As indicated earlier, Sukhomajri's approach to community integrated watershed management has become a model for watershed development programmes elsewhere in India, pioneered by both governmental and non-governmental organizations. Professional Assistance for Development (PRADAN) (whose founder, Deep Joshi, also played a significant role in the initial development of the Sukhomajri programme), Samaj Pragati Sahayog (SPS) in Madya Pradesh and the Watershed Organisation Trust (WOTR) in Maharashtra (see Lobo, 2005) are among the many NGOs that have embarked on significant watershed management programmes in other parts of India. These groups have achieved impressive results. SPS, for example, whose work is based on the principle that 'watershed development is not just about harvesting rainwater but sharing it equitably and managing it collectively', claims that in the areas in which it is working it has achieved a '4-fold increase in irrigated area, 500 per cent rise in farm incomes, doubling of land productivity and a near complete end to distress migration' (Samaj Pragati Sahayog, 1999). An in-depth analysis of the impacts of watershed management programmes in India by Joshi et al (2005), which evaluated 311 case studies of watershed programmes in terms of economic efficiency, equity and sustainability (sustainability was measured in terms of irrigated area, cropping intensity, rate of runoff and soil loss), showed that these programmes yielded an average internal rate of return of 22 per cent, with the highest benefits coming from watersheds where people's participation was highest.

Lessons learned

The 30-year experience with Sukhomajri yields numerous lessons that are relevant not only to other watershed management initiatives, but also to integrated water resources management efforts more generally.

One key lesson is the huge importance of economic returns – and of water as the key input to generate those economic returns. In Sukhomajri, improved water access was *the* catalyst for sustainable watershed development. Water was the key to encouraging people to invest in environmental sustainability. Seckler (1986: 1020) notes that the initial success of the Sukhomajri effort

was the result of the simple fact that irrigation in the arid environment of the Shivaliks was able to deliver spectacular returns in a short period of time, and thus provided a strong incentive for all parties to agree to implement important technical and social changes. Soil and water conservation in the upper catchments would not have been successful if it had not addressed first and foremost the livelihood needs and concerns of catchment inhabitants.

A related lesson is the need to pay attention to ensuring visible and substantial benefits and manageable transaction costs, a point emphasized by Kerr (2002). Kerr notes that in the Sukhomajri case, the benefits to both Sukhna Lake and Sukhomajri watersheds were substantial and clearly attributable to watershed protection, and the transaction costs involved in sharing benefits and self-regulation in Sukhomajri were manageable, perhaps because of the unusual degree of homogeneity and cohesiveness within the village.

Second, the Sukhomajri case makes clear that when productivity, conservation and equity objectives are in harmony with one another, the results can be synergistic and far-reaching. In Sukhomajri, the initial availability of water provided not only an opportunity for increased crop production and incomes, but also a strong incentive for villagers to graze their animals outside of the watershed. In addition, the system of water allocation that benefited villagers equally ensured that all villagers had such incentives, and not only those owning agricultural land. By the same token, the latter years of the Sukhomajri case also show that when productivity, conservation and equity objectives get out of harmony with one another, a downward spiral can occur. The gradual introduction of private tubewells for irrigation in Sukhomajri, while perhaps increasing crop production for individual farmers, led to a moving away from the system of equitable water distribution and a reduction in incentives for watershed conservation.

Third, Sukhomajri highlights the importance of effective legal arrangements in making equitable management arrangements work and for ensuring that the rights of the weak are protected. Seckler (1986: 1020) notes that every member of Sukhomajri's initial water users association was required to sign a legally binding document that spelled out their agreement to various provisions regarding water distribution and grazing rights, and which provided a basis for legal intervention by the appropriate authorities if needed. Kerr notes that Sukhomajri's landless people, though politically weak, were nevertheless able to exert leverage over the landowning population since, as livestock owners, they could allow their livestock to overgraze the watershed and therefore threaten much needed irrigation supplies.

A fourth important lesson is that institutional innovation often needs to go hand in hand with technical innovation. Seckler (1986: 1017) observes that, in the Sukhomajri case, the technical innovation to convey water through buried pipes made possible the equitable water distribution system that became the programme's key institutional innovation, since it enabled farmers to know reliably how much water they were getting (since time allocation was directly related to quantity) and reduced their temptation to divert more water than they were entitled to.

A fifth lesson relates to the importance of 'integrating across the time dimension' (i.e. trying to anticipate possible second-generation problems). In the Sukhomajri case, the possibility that the existence of the dam would lead to a future chain of events with adverse consequences – higher groundwater tables, shallow tubewells being dug, reduced incentives for controlling watershed grazing and participating in communal activities, overexploitation of the groundwater resource – does not seem to have been contemplated in the early stages. The failure to anticipate possible second-generation problems has been at the core of many of the world's most serious water-related environment and development disasters worldwide, from arsenic in Bangladesh's drinking water to the drying up of the Aral Sea. To ensure sustainability,

particularly in the face of climate change, two features of water resource management are critical: integrating across the time dimension and putting in place institutional arrangements at different levels that are able to deal effectively with changing conditions.

Finally, the Sukhomajri case reinforces the need for vertical as well as horizontal integration (i.e. for actions at one level to be supported by actions at other levels). While the focus of the Sukhomajri programme was at the watershed level, much of its success was due to the supportive actions undertaken at sub-watershed levels, by individuals and households as well as by the community as a whole. As importantly, Sukhomajri's initial success was no doubt linked to the supportive role played by two national and international agencies (including the Ford Foundation) in the initial stages of the Sukhomajri programme, which not only provided financial and technical support but also acted as impartial arbitrators to resolve conflicts and develop water-sharing arrangements (Kerr, 2002; Seckler, 2008). In addition, the Forest Department provided important enabling conditions in the initial stages of the Sukhomajri programme that allowed the HRMS to flourish. But when the same Department started taxing the society's profits from the sale of bhabbar and fodder, the HRMS's revenues fell sharply and its role became much diminished.

Action at higher levels is of course crucially important when it comes to scaling up small, integrated projects. Although the benefits of the Sukhomajri programme were well documented and widely disseminated in the initial years of the programme, the extension of the approach to other watersheds took time to develop. Seckler (1986: 1021) notes that the development of large numbers of small projects of this nature requires a strong managerial system; furthermore, watershed management programmes (like most IWRM initiatives) are interdisciplinary projects that fall between the cracks of different sectoral agencies responsible for irrigation, agriculture and conservation.

In sharing these lessons learned, we have one final reflection, which is that the Sukhomajri programme was never conceived as 'an IWRM project' and never had an integrated approach as an objective. Rather, the integrated approach became necessary to address the nexus of environment/poverty problems faced in the watershed and related downstream areas. It was the nature of the problem that required a holistic approach – one that took into account, as emphasized by Narain and Agarwal (2002), all the resources of a village, from grazing, tree, crop and forest lands to water systems to livestock. It is precisely because the problem at hand dictated the integrated approach, rather than the other way around, that the lessons learned from Sukhomajri are so valuable in an IWRM context.

Note

1. The authors would like to acknowledge, with thanks, the very helpful comments and suggestions received from M.R. Khurana and David Seckler.

References

Agarwal, A. and Narain, S. (1997) (eds) *Dying Wisdom, Rise, Fall and Potential of India's Traditional Water Harvesting systems*, New Delhi, Centre for Science and Environment

Agarwal, A. and Narain, S. (1999) *Community and Household Water Management: The Key to Environmental Regeneration and Poverty Alleviation*, New Delhi, Centre for Science and Environment

CSE (Centre for Science and Environment) (1994) 'Partners in Prosperity', *Down to Earth*, 3 (15 February 1994)

CSE (Centre for Science and Environment) (1998) 'Sukhomajri at the crossroads', *Down to Earth*, 7 (15 December 1998), pp29–36

CSE (Centre for Science and Environment) (2002) 'Foisting Failure', *Down to Earth*, 11 (31 August 2002), pp1–2

CSE (Centre for Science and Environment) (2007) 'Saga of two villages', *Down to Earth*, 16 (15 November 2007), pp1–7

Franda, M. (1981) 'Conservation, water and human development at Sukhomajri', *American Universities Field Staff Report*, Hanover, N.H., USA. Wheelock House, P.O. Box 150, Hanover 03755, USA

Joshi, P.K., Jha, A.K., Wani, S.P., Joshi, L. and Shiyani, R.L. (2005) *Meta-Analysis to Assess Impact of Watershed Programme and People's Participation*. Research Report 8 to the Comprehensive Assessment of Water Management in Agriculture, Colombo, Comprehensive Assessment Secretariat

Kerr, J. (2002) 'Sharing the Benefits of Watershed Management in Sukhomajri, India', in S. Pagiola (ed.), *Selling Forest Environmental Services: Market-based Mechanisms for Conservation and Development*, London, Earthscan

Khurana, M.R. (2005) 'Common Property Resources, People's Participation and Sustainable Development: A Study of Sukhomajri', *Panjab University Research Journal (Arts)*, XXXII (18.2)

Kumar, Chetan (2003) 'Institutional Reforms in Joint Forest Management: reflecting on experience of Haryana Shivaliks', Conference Paper, Politics of the Commons: Articulating Development and Strengthening Local Practices, 11–14 July 2003, Chiang Mai, Thailand

Lobo, C. (2005) *Reducing Rent Seeking and Dissipative Payments: Introducing Accountability Mechanisms in Watershed Development Programmes in India*. Conference proceedings of the World Water Week conference held in Stockholm in August 2005

Narain, S. and Agarwal, A. (2002) *Harvesting the Rain: Fighting Ecological Poverty through Participatory Democracy*, New Delhi, Centre for Science and the Environment

Samaj Pragati Sahayog (1999) *Samaj Pragati Sahayog: An Introduction*, Madhya Pradesh, Samaj Pragati Sahayog

Seckler, D. (1986) 'Institutionalism and Agricultural Development in India', *Journal of Economic Issues*, vol. 20, issue 4, pp1011–1017

Seckler, D. (2008) Personal communication

World Bank (2007) 'Watershed Management Approaches, Policies and Operations: Lessons for Scaling Up'. A report by the Energy, Transport and Water Department, World Bank, Washington, DC.

3

A Tale of Two Cities: Meeting Urban Water Demands through Sustainable Groundwater Management

Mogens Dyhr-Nielsen

In the 1970s and 1980s, the Danish cities of Aarhus and Aalborg faced similar urban water supply challenges – their urban water demand was growing and the groundwater resources that their water supplies depended upon were under threat. In the case of Aarhus the main problem was overpumping – resulting in sulphate contamination, reduced river flows and other ecosystem impacts unacceptable to an increasingly environmentally conscious populace. In Aalborg the main problem was contamination of surrounding well fields by fertilizers and pesticides.

Resolving these issues took a supportive national policy and institutional framework – which was provided by Denmark's National Aquatic Environment Plan (NAEP) – regional planning and monitoring, and collaboration at the local level between various sectors and groups. Both supply-side and demand-side tools were applied at various levels to achieve the desired results. Aarhus's solution built on national water pricing efforts to reduce demand and involved close cooperation between the water supply and water treatment sectors, as well as city planners, to achieve the necessary mutual synergies and joint interventions. Aalborg's depended on getting farmers on board and making alliances with urban planners and environmental interest groups.

The experiences of Aarhus and Aalborg demonstrate that an IWRM approach can successfully be applied to solve local problems, particularly complex problems where economic, social and environmental factors all come into play. It also shows the importance of cooperation and coordination between levels of government.

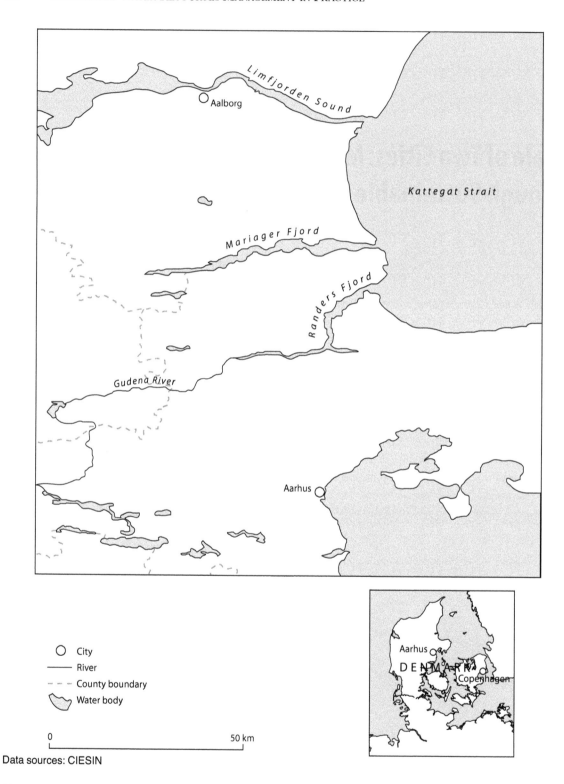

Data sources: CIESIN

Figure 3.1   Aarhus and Aalborg

The development context

During the last 100 years, Denmark has, like most of Europe, experienced significant urban and economic growth. In Aarhus, Denmark's second largest city, the population increased by a factor of five – from 50,000 in 1900 to almost 250,000 in 2005 (see Figure 3.2). During 1985–2005 alone, Aarhus had to accommodate 50,000 additional water users. During the same 20-year period, a GNP per capita increase of almost 40 per cent created significant improvements in housing standards (e.g. more bathrooms) and raised consumption potentials (e.g. large gardens with thirsty lawns). This could have led to skyrocketing per capita water demand. But in Aarhus, and many cities like it, per capita water demand, and in fact total water demand, fell during this period. This story is about the changes that made this reversal possible and the forces that drove those changes.

Population growth and urban water consumption were not the only factors that shaped Denmark's water challenges. Participation in the European Community Agricultural Policy and the resulting changes in the agricultural sector also played a role. Intensification of agricultural practices, driven by a subsidized market, led to greater use of fertilizers and pesticides, which in turn caused an increase in non-point contamination of both streams and groundwater. At the start of the 1980s, public awareness of the impact of agricultural practices on groundwater quality was minimal, and the farmers themselves were convinced that the losses of water and nutrients from their fields were negligible.

Finally, the general increase in affluence led to a popular demand for recreational areas and a clean environment. The environmental movement became strong and politically influential, and a political consensus on the importance of a clean and green environment began to emerge.

Urban water supply challenges: increasing demand and limited resources

For many years, the water supplies of Danish cities experienced relatively few challenges. Groundwater

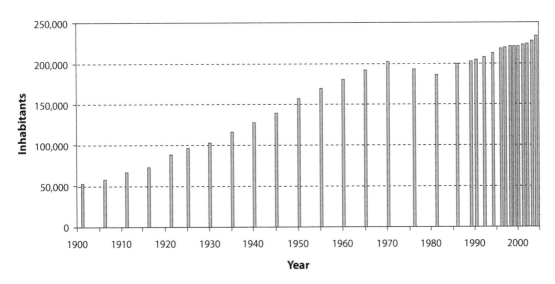

Source: Danish Centre for Urban History, University of Aarhus
www.byhistorie.dk/koebstaeder/befolkning.aspx?KoebstadID=74,

Figure 3.2 Population growth in Aarhus, 1900–2005

Source: Aarhus Municipal Water Supply

Figure 3.3   Per capita water use in Aarhus, 1900–2005 (litres per person per day)

appeared to be a safe, abundant and inexpensive resource in the first half of the 20th century, and to this day, 98 per cent of the Danish water supply is provided by the country's high-yielding groundwater aquifers.

However, in Aarhus, Aalborg and many other Danish cities, the population increased several times over from 1900 to 1970, and, as seen for Aarhus in Figure 3.3, the per capita water demand more than doubled during this period. To keep up with this demand new well fields were developed around the cities.

But, in the 1970s, several constraints emerged. Many Danish watercourses began to run dry in the summertime due to stream depletion from groundwater pumping, and the lowering water tables were having a negative impact on wetlands. The concept of environmental flow – the amount of water needed to sustain healthy ecosystems – was introduced to counteract these problems, and thus pumping intensity was reduced to achieve acceptable low-flow impacts on the biological quality of streams and wetlands.

Cases of toxic contamination were discovered, in particular in older well fields located in urban areas. Leakage from oil tanks and spills from industries using toxic materials were found in important well fields. In open farmlands the situation was not much better; many aquifers proved to be contaminated with nitrates – to such a degree that they did not meet EU drinking water standards. Also, intensive pumping had caused salinity intrusion and set off adverse chemical reactions in many aquifers, releasing contaminants such as sulphates. It became clear that remediation of contaminated aquifers was – and still is – an expensive and slow intervention.

Thus, since the early 1970s the urban water supply sector has had to face two major and closely related challenges:

- How to manage supply to reduce environmental impacts and protect groundwater from contamination.
- How to manage demand to reduce the need for expansion of the well fields. (If the per capita consumption had continued to grow at the same rate as it did in the late 1960s, a city like Aarhus would today face a water demand roughly twice current levels.)

The types of tools used to address these challenges are also twofold:

- Hydro-technical tools, such as advanced hydro-geological mapping, monitoring and modelling, have been applied to understand the complex links between pumping wells, aquifers and surface water systems.
- Socioeconomic approaches have been used to create user awareness and behaviour change.

In the following sections, we will put the spotlight on two cities, Aarhus and Aalborg, to demonstrate some actual on-the-ground responses to the challenges described above.

The Danish management approach: national policy and legislation, regional planning and licensing, and local implementation

The overall approach builds on clear roles and responsibilities between three vertical levels of management: national, provincial and local.

The national level: provider of the enabling environment

At the national level, the Ministry of Natural Resources and Environment (MONRE) is the primary authority responsible for water management. Its remit also covers chemicals and waste, noise and transport, ecosystem health, agriculture, forestry and land use. MONRE provides the legal framework, the national fund allocations for water management, and the policies and strategies. It also defines the institutional framework within which provincial and local levels operate. In an administrative reform in the early 1970s, many functions were decentralized to the provincial and local level.

A key policy tool in the water sector has been the National Aquatic Environment Plans (NAEP) 1–3, which since 1987 have provided the macro setting for Danish water management. The NAEPs apply an integrated institutional approach and a holistic view of the entire water cycle – from when the rain hits the soil until it reaches the coastal waters. From the beginning, they also included consideration of ecosystem quality and fisheries – a pioneering example of an ecosystem approach.

The NAEPs and associated water policies targeted, among other things, nutrient emissions from urban, industrial and agricultural sources. Wastewater discharges were controlled by extensive expansion of tertiary treatment in municipalities and industries. Also, agricultural nutrient losses were reduced by improved management of organic fertilizers, as well as improved land use and cropping strategies. The subsequent national outcomes have been significant, as illustrated in Figure 3.4.

During the same period, demand management via water saving campaigns and significant water price increases were implemented to reduce the pressure on the groundwater resources. The national average water price was increased 700 per cent over a 20-year period – from about 5 DKK (US$1) per cubic metre in the mid-1980s to about 35 DKK (US$7) per cubic metre today. Nowadays, an average Danish family spends approximately 4,000 DKK (US$800) a year for water, but this is still less than 1 per cent of the average annual income.

The MONRE also initiated integrated research and monitoring programmes to provide the

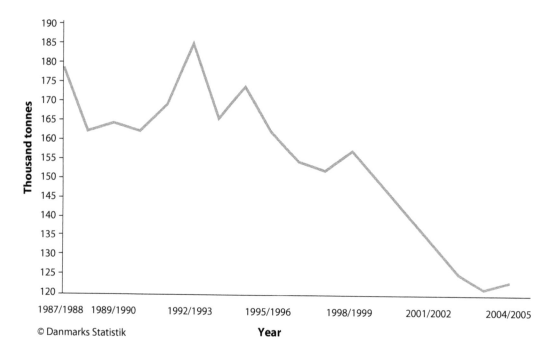

© Danmarks Statistik

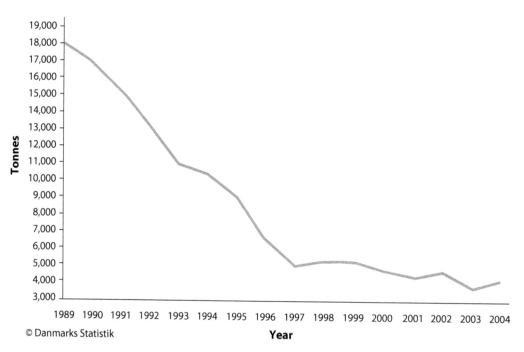

© Danmarks Statistik

Source: Statistics Denmark

Figure 3.4   Indicators for nitrate pollution from agriculture (top) and wastewater (bottom), 1989–2004

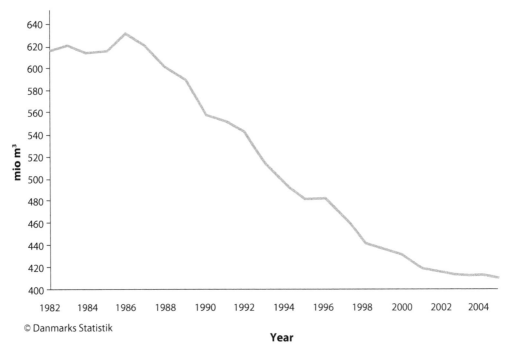

© Danmarks Statistik

Source: Statistics Denmark

Figure 3.5   National reduction in water demand, 1982–2005 (in millions of cubic metres per year)

scientific basis for proposed interventions. And it promoted institutional cooperation in sectors such as water supply, wastewater treatment, biodiversity conservation and agricultural practices.

The provincial level: responsible for regional monitoring, planning and administration

Denmark never established formal river basin organizations (RBOs), and therefore represents a case where application of IWRM concepts has been successful without the RBO institutional framework.

Up until a 2007 administrative reform, the provincial administrations were responsible for regional water management within provincial boundaries. Each province (actually counties or *amter* in Danish) had its own elected political councils and administrations, raised provincial taxes, and operated water management units covering water supply, water quality protection and wetlands conservation. Basin management issues that crossed administrative boundaries were handled via ad hoc collaboration between involved provinces and the political councils and administrations. These institutional arrangements thus avoided creating overlapping mandates between provincial authorities and RBOs.

The provincial administrations established regional development plans and also implemented monitoring programmes on the status of water resources and trends. They issued permits for water supply rights and wastewater discharges but were not involved in actual implementation, which was handled by the municipal water services as well as a large number of private waterworks.

This institutional framework encouraged close cooperation between the provincial and local levels. Through detailed knowledge of the local conditions and through close interaction between the two levels it was possible to address local challenges in a coherent and responsive manner.

The local level: responsible for public services for water supply and wastewater treatment

The reductions in water pollution and water demand (shown in Figures 3.4 and 3.5) have only been possible because of local-level efforts. The national and provincial levels are necessary *facilitators*, but without efficient local *implementers* no on-the-ground action and impact can result.

In the Danish case, locally elected municipal boards are responsible for the implementation of public services such as water supply and wastewater treatment. The municipal public works (waterworks, treatment plants, and so on) are operated on the basis of cost recovery with no access to tax subsidies. This allows for a high degree of responsiveness to local contexts, as well as economic sustainability and efficiency.

To demonstrate how the national and regional water supply management framework has been implemented in specific local contexts, two cases are presented:

- The case of the Beder waterworks in Aarhus is an example of unsustainable exploitation of a groundwater aquifer. A steady lowering of the water table, as well as water quality deterioration and environmental impacts on streams in the area, had to be addressed.
- The case of Drastrup waterworks in Aalborg is an example of a high-yielding groundwater aquifer with extreme vulnerability to contamination by nitrate leakage from agriculture.

Aarhus: balancing water demand with sustainability and environmental requirements

The context and challenges

The Beder waterworks is an important component of the Aarhus municipal water supply. The groundwater aquifer for Beder waterworks is located in the Giber river basin. In this area, aquifers are relatively small and located within a complex system of buried glacial valleys with alluvial sands enclosed in tertiary clay deposits. In the Beder case, an important outflow is through a spring that feeds the Giber River. This spring serves as a major source of flow into the Giber during the summer and other low flow periods.

Beder waterworks was originally designed for a capacity of 5.5 million cubic metres per year. However, as seen from Figure 3.6, the water table dropped by 15 metres during the period 1970–1990. Obviously, the pumping did not match the recharge.

Because of this overpumping, two serious environmental issues were emerging:

- The lowering of the water table caused oxidation of pyrite deposits and released increasing levels of sulphate into the well water, as seen from Figure 3.6.
- The springs feeding the Giber River in the dry season also ran dry, so that the low-flow discharge only comprised wastewater discharges from the treatment plants in the basin. This caused considerable political concern, since the environmental movement was growing and the demand for recreational areas for the urban population became an important voter issue.

So the challenge was twofold:

- A need to reduce groundwater pumping to match the aquifer recharge in order to stop the decline of the water table and the sulphate contamination.
- The maintenance of the low flow (today we would say the environmental flow) of the Giber River.

Instruments used

The challenges were addressed by a combination of demand- and supply-side management approaches.

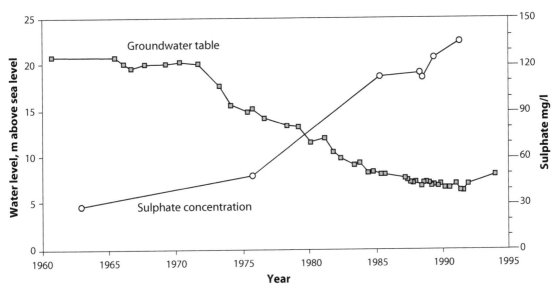

Source: Aarhus Municipal Water Supply

Figure 3.6   Water table decline and sulphate contamination

Demand-side management for reduction of water demand

At the time of its construction in 1970 the Beder waterworks accounted for about 20 per cent of the total Aarhus municipal water supply. Water consumption was then about 340 litres per person per day, with a growth rate of about 2 per cent per year. If demand had continued to grow at this rate, Aarhus would today need a water supply capacity of about 50 million cubic metres, double the total capacity in 1970. Accordingly, it became both economically and technically attractive to focus on management interventions to reduce water demand.

As seen in Figure 3.3 as well as Figure 3.7, this approach has been very successful: consumption has been reduced from over 300 litres per person per day to less than 200, and the total demand has been reduced to 18 million cubic metres per year – 7 million cubic metres per year *less* than in 1970.

This reduction in demand is largely due to a combination of awareness campaigns and a steady increase in water prices, as shown in Figure 3.7. Over the period 1986–1995 the price of water was

increased by 300 per cent – from about 7 DKK (US$1.40) per cubic metre to 21 DKK (US$4.20) – and during this time the total water demand fell from 25 million cubic metres per year to 20 million. The price of water has continued to increase – in 2008 it reached almost 40 DKK (US$8) per cubic metre. Correspondingly, water demand has continued to decrease (see Figure 3.3).

The price increases were a combination of payment for water supply and wastewater treatment, a 'green' government tax and government VAT (see Table 3.1). This price structure evolved through the integration of three water issues:

- Local-level financing of municipal water supply services. This was the original water price.
- Local-level financing of municipal water treatment investments in relation to the National Aquatic Environment Plans, starting in 1986.
- National-level introduction of 'green taxes' in 1995.

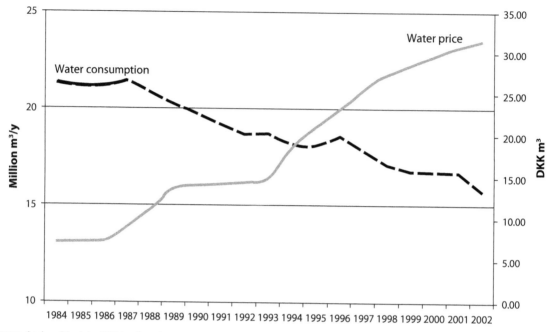

Source: Aarhus Municipal Water Supply

Figure 3.7   Relationship between water demand and water price in Aarhus

Also, the national Value Added Tax (VAT) has added to the water price. Total national taxes now amount to one-third of the total price.

The economic pricing instruments were further enhanced through local-level water-saving campaigns in Aarhus. The impact of an intensive campaign from 1989 to 1990 can be detected in Figure 3.3. Also, many technological improvements (such as water-saving toilets) were introduced during the early 2000s.

It should be noted that the impact on water demand was related to wastewater treatment charges and not to water supply charges, which are substantially lower. This collaboration between the water supply sector and the water treatment sector was a key factor in the successful reduction in the water demand.

It is also worth noting that the actual water expenditure of an average family is less than 1 per cent of annual income. It can be argued that, from

Table 3.1   *Water prices in Aarhus, DKK per cubic metre (1 DKK equals US$0.20)*

Year	Water supply	Waste-water	Water tax	VAT	Total
1985	2.85	3.79	0.00	0.63	7.27
1995	3.85	11.08	2.00	4.23	21.16
2003	6.27	13.96	5.00	6.46	32.30

Source: Aarhus Municipal Water Supply

the consumer's point of view, the price increases have hardly been noticeable. So the awareness campaigns ('good drinking water is expensive') have had a significant psychological impact on consumer behaviour.

Supply-side management via holistic analysis of the aquifer dynamics

With the success of demand-side interventions, the supply-side issues became much easier to address.

Sustainable yield

The capacity of the Beder waterworks was originally calculated using traditional hydro-geological analysis of borehole data, combined with short-term pumping tests. These methods provided an incomplete understanding of the complexity of aquifer dynamics, and as a result the capacity of the waterworks was overestimated. Thus, the falling water table and other problems due to unsustainable pumping came as a surprise when they emerged in the 1970s. Later, thanks to multi-disciplinary collaboration between national research institutes, the provincial county of Aarhus and the Aarhus municipality, advanced geophysical mapping tools (such as TEM and PATEM) were applied to obtain a more accurate description of the Beder aquifer and its complex hydro-geological characteristics (see Thomsen et al, 2004: 554). This also included development of mathematical models for three-dimensional simulation of the aquifer processes. Based on these investigations, the sustainable yield estimate was reduced by 35 per cent – from 5.5 million cubic metres per year to 3.5 million. This reduction has been readily accommodated thanks to the reduction in demand.

Low flow depletion of the Giber River

The studies also documented the depletion impacts on the low flows of the Giber River. Before pumping, the median minimum flow was estimated at about 25 litres per second. In the 1980s, minimum flow was still estimated at about 25 litres per second, but this was almost entirely made up of discharge from the wastewater treatment plants in the catchment. It was decided to introduce an artificial flow replenishment of an additional 25 litres per second. Some of this was provided from a rainwater retention reservoir, but the Beder waterworks had to reserve a river recharge capacity of 250,000 cubic metres per year, reducing the actual water supply capacity to 3.25 million cubic metres per year.

As of 2008, the Aarhus Public Works plans to close down the small treatment plants in the basin and pump the wastewater to a more efficient central treatment plant. In this case, the Beder waterworks will have to compensate for loss of the treated wastewater flows by increasing the allocation for replenishment to almost 700,000 cubic metres per year – 20 per cent of total capacity – which is roughly equivalent to the water supply demand of about 10,000 people, or a small Danish township. This is a necessary 'cost' to maintain the environmental flow and the recreational quality of the Natura 2000 site on the lower Giber River.

Vulnerability to contamination

Initially, the Beder aquifers were considered well protected against contamination due to the presence of surface layers of clay. However, the detailed geophysical surveys detected significant variability in the thickness of the clay cover, and a vulnerability map was established. This map now serves as a guide for further urban development – vulnerable areas are reserved for recreational use and urban areas located in the areas with thick clay covers.

Outcomes of the integrated approach

The management of the Beder aquifer system involved national level strategies and plans, as well as general support from the regional authorities. In the municipality itself, close cooperation between the water supply sector, the water treatment sector and the city planning sector was essential to achieve the necessary mutual synergies and joint interventions.

Thanks to the introduction of significant water price increases, combined with public awareness campaigns, it was possible to reduce per capita demand by 30 per cent from 300 to 200 litres per person per day, which allowed pumping from the Beder aquifers to be reduced to sustainable levels. A major portion of the revenue gained from the price increase was spent on investments in wastewater treatment in Aarhus.

The environmental impact on the flows of the Giber River was mitigated by artificial recharge, and the lower part of the Giber River maintains its status as an EU Natura 2000 protected area.

Also, close cooperation between water authorities, water services and the research community provided the necessary data and understanding of the complex hydrology of the basin. This detailed knowledge was used by city planners to locate recreational areas in the vulnerable zones and place urban development over the best protected aquifers.

Aalborg: protecting vulnerable aquifers against non-point agricultural contamination

Context and challenges

Drastrup waterworks is one of two major well fields in the city of Aalborg, located in the northern part of Jutland. The municipal water supply provides water to about 100,000 people, and Drastrup waterworks account for about 30 per cent of this supply, approximately 3 million cubic metres per year.

The region's aquifers are highly productive, so there are no serious shortage issues. Nevertheless, Aalborg water supply has been able to realize the same spectacular reductions in water demand as seen in Aarhus and thus achieve considerable savings in investment and operation costs.

The limestone aquifers around Aalborg are highly permeable, so the issue of sustainable yield is of minor concern. However, the aquifers are mostly unconfined, without protective clay layers. As a result they are highly vulnerable to contamination. Initially, there was an attempt to protect underlying aquifers from urban point sources of contamination such as solid waste dumps, oil tanks and gas stations. However, since aquifer protection – as well as remediation – is very difficult and costly, Aalborg water supply has preferred to locate well fields in rural areas such as Drastrup, away from the pollution-prone urban sites.

But in Denmark's rural areas, agricultural intensification has caused extensive non-point nitrate contamination of the upper layers of the unconfined aquifers, such as those around Aalborg. Figure 3.8 shows a trend in increasing contamination of the upper (or younger) groundwater layers over the period 1990–2000. The upper levels of groundwater show concentrations of 100–130 milligrams of nitrate per litre, which is more than double the EU limit value of 50 milligrams per litre.

The EU nitrate limit value may still be satisfied by mixing groundwater from the upper layers with deeper groundwater, but the situation is similar to the lowering of a water table: the resource – in terms of sufficient quality water – is being diminished. And here, a reduction in pumping will not solve the problem: leakage of agricultural nitrate from the rootzone must be reduced to protect the investments in the waterworks.

Instruments

To protect the Drastrup well fields against nitrate contamination, Aalborg water supply has applied a number of instruments, implemented within a coherent framework encompassing national, regional and local levels.

Monitoring nitrate processes in groundwater aquifers

An important component of the National Aquatic Environment Plans was the implementation of extensive, multi-disciplinary monitoring and research programmes. Initially, the agricultural sector did not accept the need to reduce nutrient losses, but the research and the monitoring data not only helped improve mutual understanding but also provided the hard facts of the extent of the contamination. The data on groundwater contamination was particularly convincing for farmers, since it became evident that their own water supply wells were the main victims. Thus, these national and regional programmes were important instruments in raising political awareness and winning acceptance of the need for interventions.

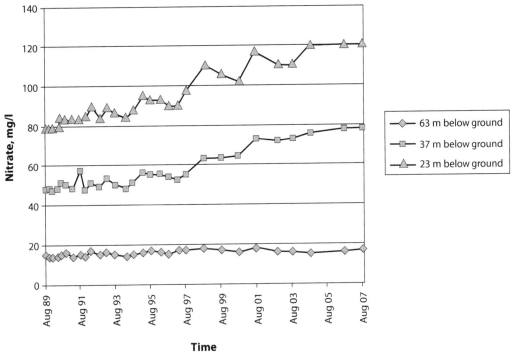

Source: Geological Survey of Denmark and Greenland (GEUS)

Figure 3.8   Nitrate levels in groundwater

Improvement of agricultural practices

Since the 1980s, the National Aquatic Environment Plans have introduced a large number of national interventions and incentives to reduce the nitrate contamination of Danish waters. These interventions address agricultural practices and have been established and implemented in close collaboration with the Ministry of Agriculture and the Danish Agricultural Advisory Service. Today, these efforts have reduced rootzone nitrate losses by 50 per cent. This reduction has had the additional benefit of improving the quality of coastal waters – a primary objective of the first NAEP. But even with the improved agricultural practices, nitrate leakage is above the limit of 50 milligrams of nitrate per litre. In many areas, this is not a critical issue, since the aquifers have sufficient natural protection, as in the Beder case. But agricultural activities near vulnerable well fields such as Drastrup pose

significant threats. It is generally agreed that the nitrate leakage cannot be reduced much further without jeopardizing the economic viability of most Danish agricultural production systems.

Land use planning

In Drastrup, land use planning has been used to address the issue of non-point contamination in agricultural production landscapes. It was proposed by the city planners in the 1980s as a tool to protect the groundwater against nitrate contamination. Combining the need for recreational areas near the city with the need to protect the well field investments established alliances and synergies between the waterworks and land use planners. It is interesting to note that, in the Drastrup case, the initial drivers of this alliance were the urban planners, not the provincial water planners. Integrating the water protection objective into their agenda enabled them to promote their own objective to establish more recreational areas.

One of the instruments laid out in the NAEPs was economic incentives for private reforestation in vulnerable water supply catchments. By 1985, Aalborg had already established a framework for the 'green' element in the city's development plans, and here the Drastrup catchment was selected as a special protection zone with multiple uses by both recreation and agriculture. Starting in 1990, the municipality tested various instruments to change the land use:

- The use of incentives to change *agricultural practices* was not a success. Farmers had very limited interest in participating.
- The use of *national* funding for reforestation projects was more successful and created visibility and political awareness of the groundwater protection issue. Nitrate leakage was not measured until the late 1990s, but it is estimated that before reforestation it was at least of the same magnitude as that found in the upper groundwater layers, namely about 120 milligrams of nitrate per litre. After reforestation, the nitrate content in water from the rootzone leakage fell close to zero.
- *Land redistribution* was also proposed, but with limited success. Unless the redistribution is voluntary, the costs – and the legal hassles – become significant.
- The inclusion of *voluntary land use restrictions* in official land titles has worked to some degree, but it is highly dependent on owner attitudes and perceptions. Some farms have agreed to change to grass with low-intensity livestock rearing. In such fields the nitrate content has decreased to a few milligrams of nitrate per litre.

In parallel with the efforts of land use planners, the Aalborg water supply established an action plan for the Drastrup catchment, based on the newly revised Danish Water Supply Act (Aalborg Kommune, 2001). This plan, which is now under implementation, provides additional resources to continue the protection efforts. However, in accordance with the mandate of the public works, it emphasizes the most cost-efficient protection of the well fields, more than the recreational interests of the entire catchment. A close coordination of the well field protection and the recreational interest groups is needed to achieve optimal results for both.

Outcomes

Forests, grasslands and meadows now dominate the catchment. But there are still substantial areas of farmland, and it has not been possible to reach a voluntary agreement with a large farm just next to the well field.

As a result of land use changes, the sources of nitrate contamination in the catchment as a whole have been reduced significantly (Møller Madsen et al, 2002). The initial impact of these measures, which were started in the late 1990s, can be seen in Figure 3.8. There appears to be some stabilization of the increases in nitrate concentration in the upper groundwater, but the concentration is still way above the EU limit value of 50 milligrams per litre. As may be expected, there are significant time lags between the reductions in nitrate leakage and the response in the groundwater quality.

Also, pesticide contamination risks have been reduced. But the small township of Frejlev is still within the catchment and creates risks for contamination from accidental spills, leaking oil tanks, and so on.

Aalborg's water supply authority is now looking for new well fields further away from urban and agricultural areas, such as in existing forests. However, forest lands – and their wetlands – may be impacted by the pumping, as in the case of Beder and the Giber River. So the efforts to find ways and means of safeguarding the existing water sources continue.

Lessons learned

It is apparent from the cases above that:

- Economic efficiency was increased substantially via increases in the water prices, coupled with public awareness efforts. By including consideration of the high cost of treatment plants, it was possible to argue for price increases to reduce water demand and thereby the need for costly water supply expansions and groundwater protection interventions.
- Equity was achieved through provision of a safe and reliable water supply for all at an affordable water price.
- Environmental sustainability was addressed by advanced methods to map and model the linked processes of the water resources and their reactions to human activities.
- At the regional level, the Danish experience has shown that IWRM approaches can be successfully applied without the presence of formal river basin organizations. Actually, the major impacts of the NAEPs has been in rivers, lakes and coastal waters.
- At the local level, each catchment had its unique issues and challenges. The actual management interventions were highly context sensitive and every case has its own story.
- Integration and cooperation were achieved by joint and specific efforts targeted at specific water challenges, where each partner had a well-defined stake. Integration is difficult if the only objective is integration for integration's sake.
- Effective water management interventions were dependent on sociopolitical contexts as much as – and probably more than – hydro-geological characteristics.

The Aarhus and Aalborg cases also demonstrate that, even at the level of local water supply issues, the implementation of truly integrated water management is a difficult long-term process that is highly dependent on:

- leadership by creative and committed water managers with ability and willingness to react to political windows of opportunity;
- multi-disciplinary and innovative research frameworks that enable cooperation around shared issues; and
- effective advocacy NGOs able to raise issues and mobilize public awareness and participation.

Such conditions will always be context sensitive and must be born out of local challenges and opportunities. No case study can provide 'the solution'. However, it is possible to learn from others' experiences. Through the Danish Water and Waste Association, there is an extensive and systematic exchange of experiences and lessons learned between the different waterworks, as well as between administrative levels. This organization has served as a very efficient tool for capacity development in Danish water management at all levels – local, regional and national.

Epilogue

The cases described above were based on a decentralized administrative structure of national, provincial and local levels, each with elected politicians with mandates to raise taxes for national, regional and local budgets. In a major administrative reform in 2007, the provincial level was removed and its former tasks were transferred to national and local levels. Also, the number of municipalities was reduced by two-thirds through mergers.

The reform has resulted in significant centralization at both national and local level. It has also created a significant loss of institutional knowledge and capability, in particular in relation to the coordination functions of the provincial and river basin levels.

There is a concern in the Danish water administrations that reform may have weakened the coherence and capabilities of water management frameworks in Denmark and may affect the country's ability to implement the EU Water Framework Directive – legislation which Denmark's water managers and experts contributed to substantially. But the reform is a result of political issues and priorities outside the water sector, and it will be interesting one day to study the impacts of this exercise on water management.

References

Aalborg Kommune (2001) 'Action plan for Groundwater Protection of the Drastrup Aquifer' (in Danish), Aalborg Kommune, Aalborg, November 2001

Møller Madsen, L., Tophøj Sørensen, M., Friis Jensen, O., Bransager, K. and Ramshøj, G. (2002) 'Evaluation and Conclusions from the Drastrup Pilot Project – an example of groundwater protection and reforestation' (in Danish), Aalborg Kommune og Skov og landskab (FSL), Hørsholm, 2002

Thomsen, R., Søndergaard, V.H. and Jørgensen, K.I. (2004) 'Hydrogeological mapping as a basis for establishing site-specific protection zones in Denmark', *Hydrogeology Journal*, vol. 12, pp550–562

4

Wetlands in Crisis: Improving Bangladesh's Wetland Ecosystems and Livelihoods of the Poor who Depend on them

Mary Renwick and Deepa Joshi[1]

The extensive wetlands of Bangladesh provide food and income to about 70 million rural people and are a critical habitat for fish, migrating birds and other wildlife species. In the last several decades, however, the wetlands have undergone a steady decline – due primarily to agriculture intensification in the flood plains, unsustainable fishing practices and, more recently, industrial pollution. The lack of alternative livelihoods for vast poor populations critically dependent on the wetlands and the concentration of access and management authority among a handful of elites made it unlikely that these trends would be reversed without outside intervention.

The Management of Aquatic Ecosystems through Community Husbandry (MACH) project, which ran from 1998 to 2007, has not managed to resolve all of these issues; however, it has been able to achieve greater positive impacts than previous interventions. Some might argue that the secret to this relative success was resources – the project had US$14 million to spend over nine years. But while the money certainly helped, several other characteristics distinguished MACH from its predecessors. It looked at the wetlands as a system – one in which there are intricate connections between land, water, people and other living organisms. It involved all the actors who impacted the system – the resource users and the polluters – and attempted to take into account the differing priorities of these actors. And finally it worked to ground natural resource management in the community, while simultaneously fostering links to higher levels of government.

MACH shows that, at the local level, good management often addresses both land and water resources, institutional as well as infrastructural interventions, and vertical as well as horizontal linkages. The case also shows that for pilot efforts to have broader impacts, effective monitoring is critical.

Water, wetlands and development in Bangladesh

The floodplains of Bangladesh form one of the world's most important wetlands, providing habitat for a vital inland fishery, aquatic plants, migrating birds and other wildlife. An estimated 70 million rural poor depend on the wetlands resources for income and nutrition. The wetlands also play an important role in the country's water resource management system by providing vital environmental services for flood control, pollution abatement and groundwater recharge. Thus, in addition to putting the health and livelihoods of millions of people at risk, degradation of these wetlands endangers many of the critical water management functions that nature currently provides for free, and which otherwise would require significant investment in infrastructure and management.

With extensive rivers and floodplain wetlands, over half of Bangladesh can be termed as wetlands; much of the country's population and wildlife is attuned to the flood pulse of the Ganges–Brahmaputra delta. Each year the monsoon rains inundate 4 million hectares of land. Fish populations disperse during this time, making them readily accessible to an estimated 80 per cent of poor rural households. As the waters recede, people cultivate paddy rice in the wet, nutrient-rich soil. In the dry season, the floodplain wetlands shrink to form a system of interconnected rivers and lakes (Ali, 1997). This is a critical time for fish reproduction; the smaller number of water bodies concentrate fish populations, supporting rapid reproduction that enable populations to regenerate.

Unfortunately, because of unsustainable fishing practices (particularly during the dry season), resource degradation and loss of ecosystem connectivity, Bangladesh's vital natural fishery is under severe threat. Since the late 1980s, the country has experienced a steady decline in freshwater fish production with serious impacts on food security, biodiversity and ecological sustainability. According to the World Conservation Union (IUCN), more than 40 per cent of freshwater fish species are threatened with national extinction in Bangladesh (IUCN, 2004). Since 1985, the national catch of major carp and catfish species decreased by 50 per cent as a result of lower productivity (Ali, 1997). The poor have been impacted the most. While national fish consumption decreased by 11 per cent over this period, it fell by 38 per cent for the poorest households (Muir, 2003).

What factors caused the decline in fish productivity and consumption in an area endowed annually with abundant water supplies and a once-vibrant interconnected system of wetlands? Several independent reviews (BCAS, 2001; Sultana, P. and Thompson, P., 2004; World Bank, 2004; Hossain et al, 2006) have identified serious underlying flaws in the management of wetlands in Bangladesh, including the following:

- **Viewing wetlands as wastelands rather than a vital part of the country's water management system.** In the past, government policies and programmes actively sought to recover land through construction of flood embankments, water control structures, and by draining wetlands for paddy cultivation. These practices drastically reduced the number of water bodies in the wetlands and impaired their connectivity, blocking fish migration routes and impacting water quantity and quality. In addition, clearing local forests to increase paddy cultivation has resulted in high rates of soil erosion, which in turn increased siltation, thereby decreasing storage capacity of water bodies. The failure to address industrial pollution problems, particularly those associated with textile dyeing near Dhaka, has resulted in serious short- and long-term impacts on fish breeding and on the entire wetlands.

- **Attempting to maximize government revenues through short-term leasing of fishing rights in permanent water bodies.** Most permanent water bodies in Bangladesh are government property. Policies have sought to maximize government revenues from permanent water bodies by awarding short-term fishing rights through leases to the highest bidder. This policy has concentrated returns from fisheries into the hands of powerful leaseholders, who often engage in destructive and unsustainable fishing practices to maximize short-term income. This policy has edged out poor fishers who are economically and nutritionally dependent on fisheries, increased competition in fishing practices and has created incentives for commercial and local fishers to over-harvest fish, particularly in permanent water bodies.

While many of the problems confronting the wetlands have long been recognized, little has been done to reverse this trend. Despite good intentions, past attempts to address wetlands-related problems have been ineffective for two principal reasons. First, most interventions have focused on a small subset of issues impacting the wetlands, such as fisheries or paddy production. This narrow approach ignored the interconnected relationships, including ecological resources, users, and rules and norms governing wetlands use. Second, past management approaches have lacked effective vertical integration; they were either top-down or entirely local. Top-down management approaches involved the imposition of best-practices on local users without taking into account specific local needs. As a result, top-down approaches rarely got community buy-in and compliance with management plans. In contrast, while community-based management approaches often got significant local buy-in, they lacked long-term support from local government, who often reverted to past practices once the project withdrew support.

Genesis of the MACH project

The nine-year, US$14 million, two-phase MACH project was jointly conceived and funded by the government of Bangladesh and USAID. MACH's objective was to be a testing ground for community-led natural resource management, with field operations in more than 110 rural fishing villages (WRI, 2008). Winrock International, an international NGO specializing in sustainable resource management projects, devised the institutional arrangements and undertook programme management in collaboration with the Bangladesh-based Center for Natural Resource Studies, Caritas Bangladesh and the Bangladesh Center for Advanced Studies (WRI, 2008). From the beginning the programme adopted a broad integrated approach that recognized the interconnected relationships affecting wetlands and included vertical as well as horizontal links, thus attempting to overcome the difficulties associated with the narrow approaches that had characterized past efforts.

Since the 1990s, government agencies, international organizations and local NGOs recognized the need to collaborate (Hossain et al, 2006). The MACH project sought to establish greater community participation through local resource management institutions and unlike most other initiatives, which focused only on building consensus among fishers' groups, the MACH project distinctly identified all the groups which depended on the wetlands (economically and nutritionally) and impacted the resource. By bringing divergent interests together, the project increased the scope for negotiation and reduced conflict. MACH targeted poorer groups, but included local elites as well to ensure sustainability. Another key to MACH's success was that it emphasized that conservation cannot be sustainably achieved in the absence of livelihood security.

The critical challenge for the MACH project was to achieve these goals in ways that were socially equitable, economically efficient and sustainable

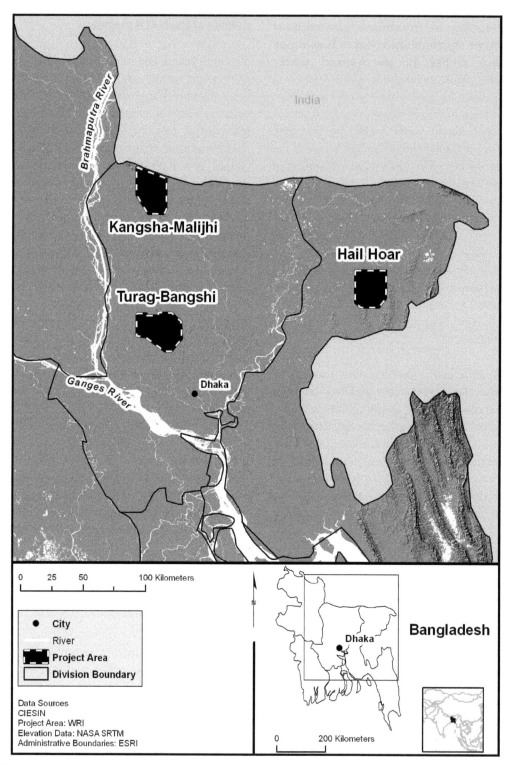

Location map

Source: WRI, 2008

Figure 4.1 The MACH project area

over the long term. The project focused on restoring and then maintaining enhanced biodiversity and production of floodplain habitats, taking into account land, water, fish, vegetation, wildlife, agriculture and other resources. Conservation and management interventions were identified by the local people and included a range of livelihood priorities and concerns:

- the rehabilitation of degraded water bodies and restoration of water corridors, primarily to create perennial habitats, essential for the biological cycles of wetland plants and animals;
- the restoration of swamp forests with local tree species, forest access and use rights to local communities;
- the reduction of soil erosion and stabilizing of river and stream banks;
- the restocking of locally threatened fish species including restrictions on use of harmful fishing gear and practices, and protection of migratory birds;
- the creation of wetland sanctuaries with community established regulations on access and use of resources; and
- capacity building and financing to promote alternative livelihoods for the resource poor, to reduce pressure on wetlands resources.

The project area includes three large and distinct wetland ecosystems (Figure 4.1), covering about 32,000 hectares and almost 700,000 people. Collectively, these wetlands provided a representative picture of the wetland habitats in Bangladesh and the poor who depend on them (see Table 4.1).

Approach

The MACH project was based on a holistic view of the entire wetlands ecosystem and a co-management approach with shared responsibilities between diverse users and local institutions in

Table 4.1 *Characteristics of the three wetland ecosystems in the MACH project*

	Hail Haor	Turag-Bangshi	Kangsha-Malijhi
Characteristics	Large, deeply flooded basin in northeast Bangladesh	Typical low-lying floodplain close to Dhaka in central Bangladesh	Flashflood-prone basin system in Sherpur district bordering the hills in India
Wet-season area (ha)	14,000	10,000	8,000
Dry-season area (ha)	3,000	700	900
Population	172,000	225,000	279,000
Number of villages	61	226	163

determining and implementing norms and rules relating to wetlands resource access, use and conservation. Additionally, the project recognized that there couldn't be a blanket solution: project staff considered the physical characteristics of the wetlands, the settlement of communities around the resource, pre-existing property rights (such as leases) to the wetlands and the social characteristics of the user, resulting in design and community organization specific to different sites.

The key elements of the MACH approach included:

* working with communities to build equitable institutions for sustainable use of wetland resources that represented all stakeholders;
* formally linking these organizations within the existing local government system;
* building the capacity of institutions; and
* working through these institutions to identify problems and solutions to restore and enhance wetlands productivity and improve livelihoods.

USAID specifically required that 'local communities have direct control over the management, utilization and benefits of local resources' (MACH, 2006). However, unlike most community-based approaches, where responsibilities for management lies largely with the community, co-management approaches seek to develop vertical linkages between communities and the government at the local, intermediate and national levels. Achieving effective and sustainable vertical linkages in the socioeconomic context of Bangladesh required major changes in institutions, organizations and attitudes.

To spur change, MACH emphasized developing equitable local institutions and supporting changes in attitudes and practices among users and local government agencies. Given the range of livelihoods practised, heterogeneity in poverty of dependent populations, and the multiplicity of actors and agencies responsible for regulating access and use of resources, it was essential for the project

to enable flexibility in developing area-specific programmes. This approach meant MACH was process based, and demanded significant resources and time, as well as requiring a higher level of capacity building in field-based staff.

Wetlands stakeholders engaged in problem identification, project planning, implementation, management and monitoring. The adaptive, open and flexible management approach enabled equity by specifically targeting the needs and concerns of the poorest in devising government-approved plans with local stakeholders and implementing them with the support of project staff and funds.

MACH helped develop interacting organizations that became the institutional basis for co-managed, community-based organizations and local government committees. Two types of community-based organizations were formed: Resource Management Organizations (RMOs) for resource management purposes, and Resource User Groups (RUGs) for livelihood development. These groups were then linked to local government through the formation of Local Government Committees (LGCs), which consisted of officials, elected representatives and leaders from community-based organizations. Emphasis was placed on making these institutions self-reliant and self-sustaining, and on establishing transparent procedures for accountability.

The creation of institutional linkages between community-based organization and local government, coupled with development of a common commitment to wetlands restoration, were important parts of the project. In particular, linkages between line agencies (Upazila administration) and elected local government (Union Parishad) through formal government orders ensured legal acknowledgement of these institutions and therefore their effective functioning. Positive reforms in wetlands use policies and practices stimulated an efficient linking of local government with organized community-based organizations.

Significant time and flexibility allowed organizations to develop and establish institutional relationships with relevant authorities, specific to each location. For example, in each location, management activities were not designed until the second year of the project when communities and local governments had collaboratively identified problems and potential solutions. Finally, stakeholders were not just given responsibilities to plan or implement; they were involved in all steps of the process including access to financial resources and the creation of an enabling institutional environment that allowed them to achieve what they had planned. This helped create ownership of the project at both the community and local government levels.

Ecosystem restoration and protection

Communities identified that restoration of the wetlands included desilting water bodies, reforestation, reintroduction of locally lost and threatened fish species, sanctuary creation, and the establishment of closed fishing seasons and fishery norms (including changes in leasing of public waters).

Three interconnected outcomes were desired – ecosystem restoration, equity and social empowerment, and economic development. In addition, MACH enabled a planned phase-out of project interventions, including technical and financial support for satisfactory completion of critical habitat restoration in project sites and support for institutional outreach. This included endowment funds transferred to relevant co-management institutions to sustain management activities.

Improved management of open-water resource

RMOs identified water bodies within their respective wetlands management areas that were strategic to fish breeding and had been affected by siltation such that they could not support fish in the dry season. To help restore the wetlands, canals and low-lying water bodies were re-excavated to improve water flows and water storage. During the course of the four-year project, RMOs oversaw the excavation work and managed contracts involving 46 hectares of lakes and 30 kilometres of interlinking canals.

Reforestation to reduce erosion rates and future siltation of water bodies

RMOs identified that land reclamation for agriculture had resulted in a rapid loss of native swamp tree species, which increased erosion rates and water-body siltation. One proposed solution was reforestation initiatives to increase tree cover and check soil erosion. Over 600,000 saplings of 56 species (48 native and eight domesticated exotic) were planted over five years. Around 40 per cent of the planted trees are expected to survive, resulting in about a 20 per cent increase in tree cover. By 2021, the net value of these trees is estimated to be about US$1.5 million, excluding potential environmental benefits. Additionally, contours were built on ecologically fragile land tracts, with commercial pineapple and other vegetation cultivated for further strengthening.

Restocking of locally lost or threatened fish species

RMOs identified key native species whose populations had declined. Project funds were used to support the local Department of Fisheries offices to restock about 1.2 million fish, mostly juveniles belonging to 15 native species.

Wetland sanctuaries

One of the most important resource management interventions was the establishment of 56 wetland sanctuaries (covering 173 hectares) by RMOs within the three wetland ecosystems. Most sanctuaries were created in water bodies where RMOs either had, or were able to secure from the government, fishing rights for five to ten years. These sanctuaries are now part of the local management plans designed to restore fish catches. A few sanctuaries were directly decreed by the Ministry of Land, following proposals made by RMOs and MACH project staff. The Ministry of Land has permanently removed these nationally important sanctuaries from the government fisheries leasing system.

Sanctuaries range from less than one hectare to over 100 hectares in size and retain water throughout the year. In most sanctuaries, the community has agreed to ban all fishing to allow breeding and repopulation during the monsoon. While the sanctuaries are primarily for restoring and enhancing fisheries, they also benefit other aquatic life including water birds and aquatic plants. This is particularly the case in the large permanent sanctuary established in Hail Haor that within two years has attracted up to 7,000 wintering water birds.

Controlled fishing in sanctuary sites and grant of water-use rights to the local community

The benefits from project supported physical interventions often prove unsustainable, if they are not complemented by policy and institutional changes. In the case of the MACH project, some resource conservation initiatives undertaken by the RMOs catalysed significant institutional changes relating to resource use and management. For example, RMOs, in close consultation with local fishers, enforced a two- to three-month fishing ban in early monsoon season when fish breed, allowing fish in the protected sanctuaries to repopulate the floodplain. Destructive fishing practices were also banned, including fishing techniques that closed off channels or captured all fish including juveniles. Water extractions from beels (lakes) for agriculture were regulated in the dry, non-monsoon seasons, to ensure maintenance of minimum water levels required to support fish life.

A major political achievement was issuing exclusive fishing rights to RMOs in some specific areas for five to ten years by the Department of Fisheries and Land. In addition, advocacy by local authorities resulted in the Department of Fisheries permanently recalling all private fishing leases in some project areas.

Industrial pollution mitigation

One of the biggest industrial clusters in Bangladesh is located in Kaliakoir, north of Dhaka, where there are many textile dyeing factories. During participatory planning processes, communities in the Turag River floodplains reported that these industries use the surrounding wetlands as a disposal ground for untreated waste, which they believed resulted in poor catches of bad smelling fish. However, they lacked evidence to substantiate their claim.

Through MACH, the community received training and equipment to monitor water quality. Regular monitoring revealed that water had biological and chemical oxygen demands more than four times higher than the national acceptable standard. These water bodies also have seasonally high pH levels and sulphide concentrations averaging 50 per cent above the national acceptable standard. MACH recommended that dyeing industries install treatment plants and the results were positive. One wastewater treatment plant has been constructed and four more are under construction. In spite of the efforts to mitigate industrial pollution, the pollution problem is worsening due to the increase in the number of textile-related factories in the area, rising from 20 to 80 in late 2005. If industrial pollution is not mitigated, then all the other positive outcomes resulting from the project will be jeopardized.

Economic outcomes

MACH's economic benefits were related to improved fisheries and wetlands productivity, as well as alternative income-generation activities that were designed to reduce pressure on wetlands resources.

Fish populations multiply, enhancing fishery productivity and consumption with benefits maximized for local wetland users

Desilting a significant tract of water bodies within each wetland ecosystem – coupled with establishment of sanctuaries, stocking and controlled fishing – had a significant impact on fishery productivity. Within the project areas, between 1999 and 2006, fish catches rose by 140 per cent, consumption went up by 52 per cent, and average daily household incomes increased by 33 per cent (Figure 4.2) (WRI, 2008). In all three areas, fish consumption (Figure 4.3) was higher than national averages in 2004. Local fishers in the MACH project sites gained US$4.7 million from higher catches in 2004 compared to incomes in

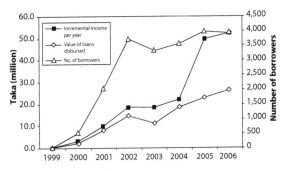

Source: MACH, 2006

Figure 4.4   Micro-credit support through MACH

1999. Increases in income allowed RMOs to repay sanctuary-related loans (Figure 4.4). Based on the success of MACH, the Department of Fisheries is replicating restocking programmes in non-project areas.

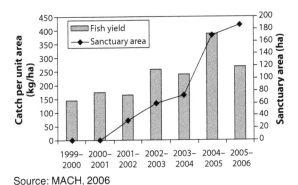

Source: MACH, 2006

Figure 4.2   Fish yield and fish sanctuaries in MACH sites

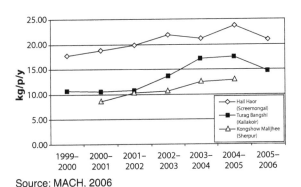

Source: MACH, 2006

Figure 4.3   Per capita fish consumption

Improved ecosystem health results in multiple food security and income benefits

Improved water management contributed to restoration of locally important aquatic plants. In the wetlands of Bangladesh, shingara has vital ecological, nutritional and economic importance. Plant stems and leaves serve as a sanctuary for plankton, help improve water quality and produce fruit that is eaten and sold locally. The plant also serves as a veritable barrier to fishing, making poaching in the sanctuaries more difficult (MACH, 2006).

In the Baila Beel in Jhenagathi Upazila (Sherpur), once-plentiful shingara was significantly impacted by extensive fishing, particularly by the use of fine mesh drag nets that remove fish hatchlings as well as aquatic plants. A 2006 MACH report (MACH, 2006) stated that:

> In six fish sanctuaries in the project area, the Shingara plant made a strong comeback, supporting the income and nutrition of around 40 to 50 poor families who collect daily four to five kg per household. Shingara sells at about 14 taka (US $0.2) per kg in local markets. In 2003, annual income from Shingara was about 7,000

taka (about US $100) per family, a significant contribution to the livelihoods of poor households.

In addition, a mid-winter waterbird census in the Baikka Beel showed a significant increase in waterbird species (Figure 4.5).

Reforestation creates multiple environmental benefits and agro-economic gains

Reforestation and contour planting have resulted in reduced run-off and erosion rates and improved soil density and fertility for farmers. With community monitoring, approximately 45 per cent of the newly planted trees have survived, and it is estimated that by 2021 the economic value of the trees will be US$4 million. Local communities retain exclusive rights to these proceeds and, with the assistance of the MACH project, communities have developed a plan for a sustainable tree-harvesting programme.

Alternative and enhanced livelihoods for poor wetland users

Poor wetland users were willing to implement the restoration plan because sound mechanisms were established to compensate for the loss of livelihood and income security associated controlled fishing and other protection activities. From the start, MACH recognized that controls on fishing – essential to reviving the productivity of wetland

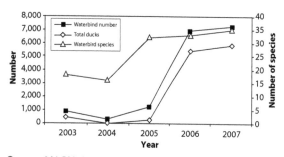

Source: MACH, 2006

Figure 4.5 Baikka Beel mid-winter waterbird census

fisheries – would critically impact poor fishers and so MACH promoted alternative income-generating opportunities.

In addition, since the inception of the project, over 5,500 of the poorest wetland resource users joined savings and credit groups through RUGs. Most of these households lived in villages close to the wetlands. Over 85 per cent of RUG households have historically depended on fishing or collecting other wetland resources for income or food and many possessed few or no assets.

Following local NGO practices for credit and savings programmes in Bangladesh, one person per household was permitted membership to a RUG. Members identified alternative livelihood opportunities and received loans on the basis of their saving records. Loans for specific enterprises were complemented with business and enterprise skill development training. Typical enterprises in the project areas include raising poultry and livestock, small shops, tailoring and tree nurseries. Over 30 per cent of members taking loans were women.

By 2005, households engaged in alternative income-generating activities had reduced their fishing effort by an estimated 20 to 30 per cent and increased their average annual income by 65 per cent compared to pre-project levels, amounting to US$0.8 million from new enterprises. A benefit–cost analysis of the MACH project, which included the value of increased fish catches, alternative income generating activities, trees and contour pineapple cultivation, showed a benefit–cost ratio of 4.7 and internal rate of return of 56 per cent (MACH, 2006).

Equity and empowerment outcomes

Given the history of elite capture of common property resources, the emphasis was on inclusion of the poorest. The project enabled sensitive negotiations between poor resource users, local government agencies, and local elites and other actors to ensure sustainability. Through this process, poor wetland users were provided with a legitimate place and voice in the project planning, management and implementation process. The motivation of the local communities to monitor and regulate conservation was high, given that these were self-developed plans, and that alternative economic options were in place to mitigate income and food security needs.

Emphasis on real participation by all stakeholders led to outcomes that were more equitable, efficient and effective

Participatory planning took different forms in different areas, but the primary goal was to engage all stakeholders in understanding the long-term consequences of environment degradation and to identify viable solutions for wetlands conservation. Workshops and discussions were held with different stakeholders (fishers, farmers and other), and separate discussions were conducted with the poorest and women in all groups. Mohammed Nuruzzaman, chair of the Bilasha RMO, says that 'Now we meet face to face with the officials and councillors, who listen to us and try to solve our problems. Now we have a status in the community, given this recognition from the administration' (MACH, 2006).

Resource conservation and restoration plans were identified collectively and were translated into implementation agendas by RMOs, in consultation and agreement with local elected councillors and government departments. During this process, it was evident that several interventions would be in conflict with existing livelihood practices of local communities. These included restrictions on practices such as intensive year-round fishing, clearing of water bodies for agricultural production, and water extractions for irrigation in dry months. To enable local households to overcome these livelihood problems, alternate economic activities, such as small-scale enterprises, livestock and poultry rearing, were promoted by the project. Interventions were targeted to the poorest households – those owning little or no land, and therefore critically dependent on wetlands resource use. Funds were provided through revolving fund mechanisms, and skills- and capacity-building initiatives were undertaken.

Reversal of fishing rights to local fishers

A notable example of the empowerment of poor local communities was the transfer of fishing rights to local fishers in Hail Haor. Prior to MACH, fishing rights were leased to private investors and middle men by the Department of Fisheries in Hail Haor. To maximize their income, private leaseholders seek to maximize short-term catches impacting on long-term sustainability of local fisheries. As a result of MACH, RMOs successfully advocated for fishing rights, and they have been restored to many local communities, especially in areas where the RMOs have been persistent and vocal.

Will economic and social outcomes be sustained over time?

A number of factors suggest that the achievements of the MACH project will be sustained within the project areas, and will possibly be replicated elsewhere.

There is evidence of increased awareness regarding the value and benefits of community-led resource conservation among local communities. An independent assessment in 2006 identified that over 92 per cent of direct project participants and around 80 per cent of the general village community were aware of RMO and RUG activities and objectives. Another critical indicator for sustained adoption of community-designed best practices was the community enforcement on fishing restrictions and use of inappropriate fishing gear and fishing in peak breeding seasons. Over 100 villages in the MACH areas and an additional 120 villages in non-project villages had adopted and regulated these practices.

Independent surveys identified that in project areas, conservation interventions were viable because the Alternative Income Generation (AIG) initiatives had consistently improved daily incomes (45 to 50 per cent gain) for RUG members after three or more years of programme support. Also 46 per cent of people adopting a new livelihood strategy had been supported by appropriate training. While more than 50 per cent of RUG members had taken loans for AIG, assessments also identified that competing priorities in the face of poverty can result in the use of AIG credit for non-AIG purposes. However, in all three project sites, the poorest households reported an increase in supplemental income and reduced reliance on fishing both in peak and lean seasons.

MACH was also able to catalyse policy reform at the national level, indicating political commitment to sustaining these changes. For example, MACH RMOs have obtained perpetual resource use rights (at no revenue fee) from the Ministry of Land in eight significant project site areas, now designated as permanent sanctuaries, and several critical water bodies in the MACH sites have been leased out to RMOs for up to ten years by the Ministry of Land. RMOs and Upazila Fisheries Committees have authorization from the Ministry of Fisheries and Livestock to function as legal entities, with significant decision-making rights at the local level.

As stated above, the MACH project established endowment funds to support two critical initiatives: alternative income-generating activities through revolving microfinance for RUG members, and support of management activities of RMOs. A number of other activities indicate that community mobilization has enabled empowerment and set the foundation for long-term sustainability. One such activity has been the active and vocal lobbying by RMOs against industrial pollution problems in the Turag–Bangshi near Dhaka. Since they learned skills and procured equipment to monitor the water quality in the Turag River and the interconnected lakes, the community groups have consistently used the resulting evidence to press government departments and industries for adopting and enforcing sound effluent-treatment practices. Their concern and activism has been picked up by numerous civil society groups, including the MACH team, the UK Department for International Development and the European Union, who have raised these issues at the national level.

Lessons learned

As highlighted earlier, the wetlands of Bangladesh play an important role in the country's water resource management system. From the point of view of this book, therefore, the most important aspect of the MACH project is that it adopted an integrated approach to prevent and if possible reverse the degradation of these wetlands.

The integrated approach adopted by the MACH project was very broad based in that it covered both land and water resources, institutional as well as infrastructural interventions, and vertical as well as horizontal linkages. Some of the specific lessons learned about the approach include the following:

- **The importance of a holistic vision.** MACH did not look at fisheries management or water management or land management, but how to manage wetlands as a *landscape* – one which is shaped by diverse biophysical, social and economic processes. This could only be done through a multifaceted, multi-disciplinary, multi-sectoral approach.

- **Co-management involves shared responsibilities and inclusiveness.** The MACH project demonstrates the multiple benefits of investing in community mobilization, capacity building of local groups and the creation of enabling environments for sustained functioning. For community organizations to function effectively, space and time needs to be allocated for a diverse range of stakeholders to collectively identify problems and develop solutions. For co-management to work, effective working relationships need to be developed between key stakeholders and government departments.

- **Participatory approaches should focus on providing all stakeholders with an equal opportunity to participate.** MACH worked with all the stakeholders, from local community fishermen, businesses, the poor and elite, local government, district government, to national-level ministries. Taking into account the priorities of both the poor and the local elite helped minimize risks of sabotage and poor governance. At the same time, the project focus on equity ensured that the poorest were targeted – through affirmative interventions and disproportional benefits. Without a concerted effort to build institutions that empower the poor, the majority of people living in poverty do not have bargaining power and do not understand their rights.

- **Involving women continues to present difficult challenges, even under favourable institutional environments.** A key drawback of the project has been the difficulty in making local organizations accessible and relevant to women. For example, independent surveys identified that while women RUG members had found the training useful, only a small number were able to apply that training to improve their incomes. Local cultural biases which serve to isolate women from public processes, including adoption of new livelihood strategies, require significant interventions.

- **An adaptive design, implementation and management approach which draws up activities as needs became apparent is required.** In the MACH project, wetland resource management plans were adapted, reviewed and approved on an annual basis according to new information and the previous year's experiences. Other examples include designing and implementing a public communications and awareness strategy, tree planting to reduce erosion, pineapple contour cultivation to reduce soil erosion, and adding a pollution abatement component to the project. Adaptive management allows for learning by doing, and openly discussing and solving challenges and constraints.

- **Creating an environment for local champions to emerge.** Through open, flexible and participatory planning and implementation processes, the MACH initiative enabled the emergence of local champions, who advocated and helped achieve significant equity and environmental gains. Their commitment has resulted in certain aspects of the programme self-replicating beyond the project boundaries.

- **Equity, environment and economic efficiency are not always in competition with one another.** The analysis of outcomes from the MACH project suggests that equity, environment conservation and economic gains are not always conflicting goals, but can be achieved simultaneously.

In addition, the MACH story illustrates the importance of effective documentation and monitoring. However, in providing evidence of MACH experiences and outcomes based on programme and independent assessments, it is important to note that, in the absence of historical data trends as well as multiplicity of development interventions in the programme sites, several outcomes may not be directly attributable to the MACH project. While results to date look promising, assessment of sustainability must be considered over a long time frame. Risks to sustainability of programme interventions as identified by project staff include the following:

- Despite coordination and uniform strategies for conservation, individual projects continue to operate guided largely by specific programme and donor requirements, often presenting conflicting options and strategies for local communities and government authorities.

- Despite policy reform, institutional memory of programme interventions and objectives at the local government department levels is often restricted to individual staff members. Frequent staff transfers present a formidable loss of project-generated awareness and capacity building.
- Donor priorities change and commitment is often not consistent to the depth and degree of time and resource investments required to achieve sustained impact at the national and local levels.
- The cost of the intensive engagement (US$14 million for 25,000 hectares) may not be replicable on a larger scale.

While these difficulties are real and it is too early to provide definitive answers, MACH project staff are confident that project practices established through institutional reform will continue. RMOs are permanent participants in the fishing planning process in local offices of the Department of Fisheries. The Upazila Fisheries Committees meet four times a year to discuss fishing plans and priorities. The Government of Bangladesh has formally recognized the Upazila Fisheries Committee instituted under MACH, and granted them the right to undertake all wetlands development work in the region. Local officials are hopeful that this process will be a model for wetlands management in all of Bangladesh.

Finally, the MACH story also provides some insights into the thorny question of upscaling pilot project results. Current results appear to have led to several concrete governance changes and replication in non-project areas. In particular, it appears that some key elements of MACH's approach have been adopted by the Government of Bangladesh in other fishing areas and in a pilot programme for community-led management of protected forest areas. In addition, MACH's co-management model has been adopted in Bangladesh's new Inland Capture Fisheries Strategy reversing the decades old policy of centralized control over the floodplains (WRI, 2008).

Note

1. The authors are on the staff of Winrock International, which devised the institutional arrangements for the project described in this chapter and was also involved in its management.

References

Ahmed, M. (1993) *Rights, Benefits and Social Justice: Keeping Common Property Freshwater Wetland Ecosystems of Bangladesh Common.* Conference proceedings of the Fourth Common Property Conference held in Manila, Philippines. Conducted by the International Center for Living Aquatic Resources Management (ICLARM)

Ali, M.Y. (1997) *Fish, Water and People*, University Press Ltd., Dhaka

Hossain, M., Islam, A., Ridgeway, S. and Matsuishi, T. (2006) 'Management of Inland Open Water Fisheries Resources of Bangladesh: Issues and Options', *Fisheries Research*, vol. 77, issue 3, pp275–284

IUCN (World Conservation Union), Bangladesh (2004) *Proceedings of the Regional Workshops on National Biodiversity Strategy and Action Plan*, held in Dhaka. Conducted by the World Conservation Union, Bangladesh

MACH (Management of Aquatic Ecosystems through Community Husbandry) (2006) Annual Report prepared by MACH for USAID

Muir, J. (ed.) (2003) *Fisheries Sector Review and Future Development: Theme Study: Economic Performance*, World Bank, Danida, USAID, FAO and DFID, Dhaka

Sultana, P. and Thompson, P. (2004) 'Methods of Consensus Building for Community Based Fisheries Management in Bangladesh and the Mekong Delta', *Agricultural Systems*, vol. 82, issue 3, pp327–353

World Bank (2004) *Bangladesh Fisheries Sector Review, Report No. 8830-BD*, Agriculture 19 Operations Division, Asia Country Department, World Bank, Dhaka

WRI (World Resources Institute) (2008) *World Resources 2008: Roots of Resilience – Growing the Wealth of the Poor*, United Nations Development Programme, United Nations Environment Programme, World Bank, World Resources Institute

5

Should Salmon Roam Free? Dam Removal on the Lower Snake River

Peter Rogers

The Snake River is a major tributary of the Columbia River in the United States. It flows from its source in Yellowstone National Park through the states of Wyoming, Idaho, Oregon and Washington. Many dams were built on the Snake River during the 20th century to provide hydropower, irrigation and navigation to fuel the region's economic development. But that economic development – the dominant priority at the time – has come at the expense of the river's salmon population, which has declined dramatically.

Now the salmon are fighting back, or rather numerous individuals and interest groups are fighting on their behalf to tear down four large dams on the lower Snake River. The efforts to resolve the conflict between the different social, economic and environmental priorities and interests involved are on-going. While the various studies, negotiations and consultations may have marginally improved things for the salmon, the dams still stand and the battle continues in the courts.

This case shows that the way water is managed must change to reflect evolving priorities, as well as shifts in the institutional and external environment. It also shows that such changes are not easy and they do not happen overnight. The case also highlights the importance of an effective legal framework to help achieve a balance among competing priorities when consensus cannot be reached.

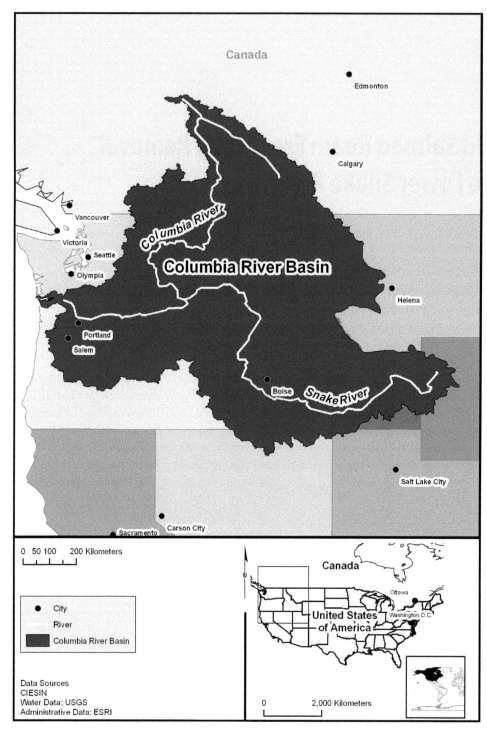

Source: US Army Corp of Engineers, 2002: 3

Figure 5.1   Map of the Snake and Columbia Rivers

Location map

From industrial to post-industrial priorities in the management of the Snake River: the water challenge

As societies develop, priorities change. Poorer societies emphasize building industry, creating jobs and expanding the economy; wealthier societies may develop different priorities. It is not that simple, however, with basin development. Many of those who benefit directly from basin development are relatively content with the status quo; the pressure to change priorities often comes from people and groups who are not the direct beneficiaries, and who may even live outside of the basin. The case of the salmon fisheries on the Snake River demonstrates how difficult accommodating these new priorities

can be, especially when it involves tradeoffs with existing beneficiaries.

The Snake River (see Figure 5.1) is typical of many North American rivers. Over the second half of the 20th century, it has been extensively developed. Now questions are being asked about the current management regime's environmental sustainability, specifically about the detrimental impact on the salmon, a federally listed endangered species. Environmental issues always concerned the planners, developers and water infrastructure managers, but other social and economic goals were higher on both their agenda and the public's. Now that many of those goals have been achieved and the environmental movement has grown stronger, the question has become how to achieve a balance among conflicting environmental, social and economic uses – the core IWRM question.

Does this new conflict mean that an IWRM approach was missing before? This author would argue no. Since the 1930s the Columbia River system's managers have tried to integrate different water uses. Even if that multiple use approach was not dignified with the name IWRM, it lay behind the building of Grand Coulee Dam, the system's first, which was finished in 1942. It was the expansion of the dam structure to include multiple uses that brought Federal and State funding into the system in the first place. Ecological flows and salmon runs were not part of the original discussion. Still, integrating multiple uses was a fundamental premise of the original planning. What this chapter addresses is not the role of IWRM in basin planning and management, but rather how an IWRM approach can help in adaptation to new conditions, in this case the newer emphasis on environmental concerns.

Developing the basin

Developing a river basin can harm its fish. Certainly the decline and disappearance of salmon from the Columbia River is not in dispute. The Columbia's was once one of the greatest salmon runs in the world, with as many as 15 million adult salmon returning to spawn each year. Current runs are only one-tenth of that (Meadows, 2004). Salmon are harmed by dams which block their passage, cause habitat loss and degradation and diminish water quality. As a result, 26 of the Pacific salmon stocks are federally listed as threatened or endangered. Although there are 24 large dams on the Columbia River and its main tributaries, the biggest controversy centres on the Snake River (Meadows, 2004), which once produced 40 per cent of the Columbia salmon run and where fishing interests claim that 90 per cent of the fish are now killed. In 2001 there was a call for breaching the four federal dams on the lower Snake River.

The Columbia River basin is home to extensive timber, irrigated agriculture and mining industries as well as major aerospace and, more recently, information-technology companies. Even though the Federal government in the 1930s supported the Columbia Basin development project as a 'conservation' activity, it was thinking of natural resource conservation, not broader environmental and ecological concerns. The reality is that the original priority was to exploit the natural resources; the dams were built to generate hydropower for the expanding industries and to create a 140-mile-long barge system. Since the 1960s there has been a growing concern nationwide about the need to protect, sustain and enjoy the natural environment, a concern reflected in national environmental legislation.

Dammed if you do, dammed if you don't

Between 1960 and 1975, four dams were built along the Snake River's lower reaches. Moving upstream they are Ice Harbor, Lower Monumental, Little Goose and Lower Granite. Figure 5.1 shows their location and Table 5.1 outlines certain technical details. These four dams provide important benefits. They generate over 3,000 megawatts of peaking power, provide transport for 3.8 million tonnes of freight including more than 40 per cent of the region's grain exports, and supply water for agriculture and municipal use.

However, the dams have also had a negative impact on fish stocks not only in the river, but also in the Pacific Ocean since the Snake River is an important spawning ground for salmon. The dams' ocean impact has recently spread to Puget Sound's killer whales, also federally listed as 'endangered', which feed on migrating salmon. When studies by the National Marine Fisheries Service in 1995 (NMFS, 1995a,b) suggested that the interruption of the spawning run was endangering a number of varieties of salmon (on the Endangered Species List) as well as other species, it was decided that action had to be taken. Besides programmes to breed and seed hatchlings, options to help the salmon get around the dams, such as fish ladders and trucking

Table 5.1 *Technical details: four Snake River dams*

	Ice Harbor Dam	Lower Monumental Dam	Little Goose Dam	Lower Granite Dam
Location (miles from confluence with Columbia river)	10	42	70	107
Year placed in service	1961	1969	1970	1975
Area of Corps-managed land (acres)	4,000+	9,100+	4,800+	9,200+
Length of reservoir (miles)	31.9	28.7	37.2	39.9
Power generation characteristics	Three 90-MW and three 110-MW generators	Six 135-MW generators	Six 135-MW generators	Six 135-MW generators
Navigation lock characteristics	90-foot-high, 86-foot-wide single-lift navigation lock	100-foot-high, 86-foot-wide single-lift navigational lock	100-foot-high, 86-foot-wide single-lift navigational lock	100-foot-high, 86-foot-wide single-lift navigational lock
Number of spillbays	10	8	8	8
Benefits	Power, recreation areas, navigation lock, wildlife habitat areas, irrigation, fish passage and port facilities	Power, recreation areas, navigation lock, wildlife habitat areas, irrigation, fish passage and port facilities	Power, recreation areas, navigation lock, wildlife habitat areas, irrigation, fish passage and port facilities	Power, recreation areas, navigation lock, wildlife habitat areas, fish passage and port facilities, municipal and industrial pump stations, port facilities

Source: US Army Corp of Engineers, 2002: 11

the juveniles downstream, were considered, as was removing the dams.

The difficulties of integrating salmon into an engineered river like the Snake are exacerbated by the sheer complexity of the life and breeding cycle of the animal. Figure 5.2 hints at this complexity and why dams are a problem for salmon migration. Without fish ladders or other forms of fish transport around the dams, the salmon are blocked from returning to their original spawning grounds. And the vagaries of turbulence and turbine kills make it harder for fish to get safely down the rivers to the ocean to repeat the cycle. This sketch also does not show the complex symbiotic relationship between the salmon and the killer whales of Puget Sound.

The approach to resolving conflicts

One cannot really understand river basin management in the USA without also understanding American federalism. The states have sovereignty over water and real economic and political power. The negotiations among the Federal government agencies – as assessed around different uses – result in a variety of cost-sharing arrangements among Federal, state, local and other entities. This negotiating system is really built from the bottom up; however, the current regulatory approach is really a Federal top-down approach. While an IWRM approach generally entails decentralization to the lowest appropriate levels (see

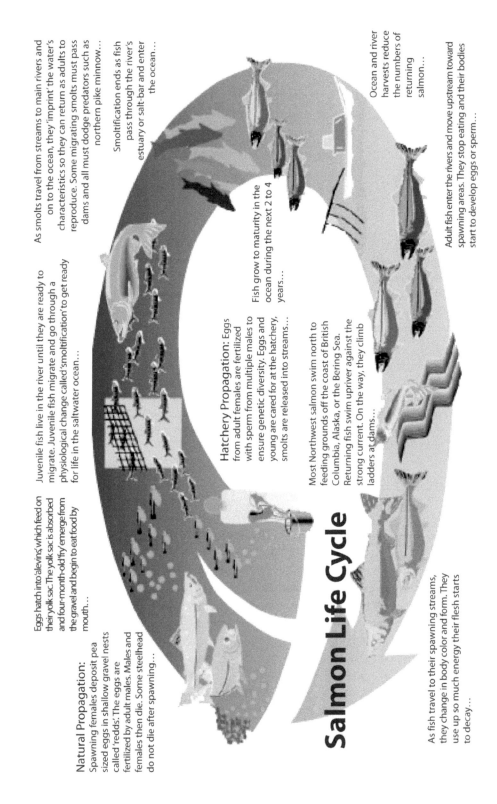

Natural Propagation:
Spawning females deposit pea sized eggs in shallow gravel nests called 'redds'. The eggs are fertilized by adult males. Males and females then die. Some steelhead do not die after spawning…

Eggs hatch into 'alevins' which feed on their yolk sac. The yolk sac is absorbed and four-month-old fry emerge from the gravel and begin to eat food by mouth…

Juvenile fish live in the river until they are ready to migrate. Juvenile fish migrate and go through a physiological change called 'smoltification' to get ready for life in the saltwater ocean…

As smolts travel from streams to main rivers and on to the ocean, they 'imprint' the water's characteristics so they can return as adults to reproduce. Some migrating smolts must pass dams and all must dodge predators such as northern pike minnow…

Smoltification ends as fish pass through the river's estuary or salt-bar and enter the ocean…

Hatchery Propagation: Eggs from adult females are fertilized with sperm from multiple males to ensure genetic diversity. Eggs and young are cared for at the hatchery, smolts are released into streams…

Fish grow to maturity in the ocean during the next 2 to 4 years…

Ocean and river harvests reduce the numbers of returning salmon…

Most Northwest salmon swim north to feeding grounds off the coast of British Columbia, Alaska, or the Bering Sea. Returning fish swim upriver against the strong current. On the way, they climb ladders at dams…

Adult fish enter the rivers and move upstream toward spawning areas. They stop eating and their bodies start to develop eggs or sperm…

As fish travel to their spawning streams, they change in body color and form. They use up so much energy their flesh starts to decay…

Salmon Life Cycle

Source: US Army Corps of Engineers, 2002: 25

Figure 5.2   Salmon life cycle

Chapter 1), the American federalist system starts from the very bottom up and asks how much Federal power can be appropriate, not the other way around. The original basin development was based upon planning begun in the early 1920s. Since the late 1970s there has been a shift in emphasis in the USA towards regulation based on several strong Federal water laws, rather than bottom-up, plan-based regulation. This shift also means that institutional structures have to evolve in step with the newer regulatory approach.

While the Snake River dams were being completed, a major legislative event affecting all water projects occurred – the passage of the Endangered Species Act (ESA) of 1973. Its passage was mostly overlooked by water managers and few realized at the time just how important the ESA would prove to be for water development in the USA. It has essentially provided a veto over developing water projects in many parts of the USA, since every water project has major impacts on fisheries and habitat. It has also sparked major scientific fights as to how precisely to define 'endangered' and 'threatened' species.

The ESA's purpose is to protect species and the ecosystems upon which they depend. The ESA forbids Federal agencies from authorizing, funding or carrying out actions that may jeopardize the continued existence of endangered or threatened species. It also forbids any government agency, corporation or citizen from harming, harassing or killing endangered animals without a permit. Moreover, once a species is listed as endangered or threatened, the ESA also requires that critical habitat be designated for it. Federal agencies are forbidden from authorizing, funding or carrying out actions that may destroy or adversely modify critical habitat. The ESA is administered by two Federal agencies, the Fish and Wildlife Service of the Department of the Interior and the National Marine Fisheries Service (NMFS) of the National Oceanic and Atmospheric Administration. It contains a citizen-suit clause that allows citizens to sue the government to enforce the law.

The ESA brings into sharp focus the need for water planning and management to adapt to changes in the physical, economic and cultural environment. Much of the US water infrastructure was planned when multiple goals were much simpler: economic development of backward regions and provision of hydropower, navigation, irrigation and domestic water supplies. More recent concerns require these same existing systems to cope with new additional goals: recreational needs, in-stream water quality, protection of wetlands and river ecosystems, in particular habitat protection for endangered species. Applying these concerns retroactively to existing management systems, such as the Snake River's, means having to give up some significant fraction of the existing benefits. If the USA were starting from scratch, higher levels of system integration might be achieved. But in the USA there are fewer and fewer new river basin projects and more and more need to adjust the existing systems to new realities. An IWRM approach in this context should aim to help managers of existing systems adapt to these new realities.

Prompted by citizens' suits brought under the ESA concerning Columbia River Chinook Salmon (fall runs), which were listed as 'endangered' in 1993, a major programme of studies and consultations involving Federal and state governments as well as environmental and fishing organizations has been under way since 1995. In December 2006, the draft of the Snake River Salmon Recovery Plan was submitted to NMFS. The proposed plan included several options for mitigating impacts on the salmon, but ignored a Federal judge's 2004 order that NMFS consider removing all four dams, if necessary, to meet the ESA's requirements.

Analysis of alternatives

Many options have been considered in analysing the question of salmon and dam removal on the lower Snake River. These included a continuation of the transport of young salmon by truck and barge, and the following four alternative development scenarios:

Alternative 1 was the 'Do nothing', or the status quo option; Alternative 2 entailed 'Maximize transport of juvenile salmon'; Alternative 3 involved 'Adaptive migration using major system improvements'; and Alternative 4 entailed removing the dams altogether or 'dam removal/dam breaching'. A cost–benefit analysis by the US Army Corps of Engineers (2002) was then carried out on each alternative. Maximum transport of juvenile salmon (Alternative 2) was the most preferred; removing the dams (Alternative 4) the least preferred.

According to the Corps of Engineers, lost hydropower benefits (US$271 million annually) was perhaps the largest negative impact of removing of the dams. On the positive side the Corps of Engineers estimated an increased benefit from recreation (US$71 million annually) even though if the dams were removed most existing recreation sites, visited by 2 million people each year, would have to be closed during the dam removal or undergo major modification. The impact on agriculture of removing the dams was found to be relatively small – about 2,300 permanent job losses in the region – reflecting a developed economy in which the economic role of farming is limited.

Not to be intimidated by the Corps of Engineers, a consortium of NGOs led by Save Our Wild Salmon (SOS) – a coalition of businesses, conservation organizations, commercial and sport-fishing groups and taxpayers' associations – produced its own economic evaluation of dam removal on the lower Snake River (SOS, 2005). It concluded: 'Dam removal will save taxpayers and Northwest rate payers between $2 billion and $5 billion over 20 years and will also generate at least $9 billion in new revenue.'

One wonders how two such radically conflicting conclusions could emerge from studies carried out on the same system. The discrepancies between the studies appear to lie in the SOS report's focus on costs, in particular the estimated cost (US$7.8–9.1 billion) of the 2004 Federal Columbia and Snake River Salmon Plan. The Corps of Engineers' study implies that by modifying the system it will be possible to meet the requirements of the 2004 Federal Plan without dam removal.

In 2007 the Northwest Power and Conservation Council's Independent Economic Analysis Board (IEAB), which advises the Council on difficult economic issues associated with its fish and wildlife programme, joined the debate. While not endorsing all of the Corps of Engineers' conclusions, the IEAB was sharply critical of the SOS report, finding fault with it on several factual and methodological points. The IEAB report essentially confirms the Corps of Engineers' economic analysis. This was later reinforced by the Bonneville Power Administration's (2007a,b) fact sheets on dam removal. This discussion appears to have clarified the economic analysis, while pointing out that the final decisions about survival of salmon on the Snake River are not likely to be determined solely by economic calculations, or mitigation activities taken on the river. The founder of an organization called Save Our Dams wryly commented: 'For every ten salmon returning to spawn, six are taken by ocean harvest. Only in the United States can you buy an endangered species for $2 a pound, continue to commercially harvest them while at the same time spend $1 billion per year to save them' (Meadows, 2004: 5).

At this stage, legal and institutional arguments have come to the fore. The revised Draft Snake River Salmon Recovery Plan (SRSRP, 2006) reports on the proposed government actions and the citizens' responses through an elaborate public participation effort. The Revised Plan itself is no more than a listing of mitigation actions to meet the specified goals. It does not recommend removing the dams.

Reconciling economic, equity and environmental objectives

In evaluating the alternatives, the Corps of Engineers measured economic efficiency with a cost–benefit methodology. It put no price on the value of protecting an endangered species, but sought only to determine the most cost-efficient way to achieve that goal. The Corps of Engineers' analysis highlighted the economic consequences of restricting the other uses of the river. The SOS study (2005), taking an opposite approach to efficient resource use, insisted on meeting, at great cost, the 2004 Federal Plan.

Social considerations were also addressed in the two analyses. A key concern was to understand the impact that proposed water management changes would have on employment. Again, there was a huge discrepancy between the Corps of Engineers' estimate of 2,300 job losses under the dam removal option compared to SOS's quoted figure of 15,000 job gains.

Notwithstanding the social and economic objectives, the main goal of the process was to find ways to reconcile the environmental needs of the endangered salmon with the other uses of the river. At the time of this chapter's writing (February 2008), no final resolution of the conflict has been reached.

Plans, plans and more plans

The effort to develop proposals for a way forward on the Snake River highlights the complexity of water management for such intensively used rivers. The process considered all significant water-based activities. It provided a framework to evaluate the impact of policy and management options. It suggested that if all sectors worked together compromises could be achieved that largely met the needs of all. However, the process also highlighted the indispensable role of the courts in achieving a balance where consensus cannot be reached – a reminder that IWRM is a governance approach that depends on an effective legal framework to be successful.

The process is still very much under way. Fish numbers appear to have improved thanks to mitigation measures such as altering the dams in the 1980s to better accommodate fish passage, and transporting young salmon around the dams. The

conclusion reached in Washington State's Salmon Recovery Plan (SRSRB, 2006) was that the best approach would be to manage the dams in a manner that would help the salmon to migrate, while allowing the other activities to continue. Washington State's plan relied on habitat and minor fish-passage improvements and did not recommend removing any dams. However, this approach was contested in 2005, when a Federal judge threw out the 2004 Federal Columbia River Power System Plan governing Federal dam operations on the Columbia and lower Snake Rivers. That plan, according to the judge, failed to adequately address recovery of endangered salmon and steelhead including the removal of the four dams on the lower Snake River. He termed the 'Biological Opinion' on which the plan was based as 'arbitrary, capricious and contrary to law'. As a result, the plan was remanded and is currently being redrafted. In a letter of 7 December 2007 the same judge excoriated the Federal agencies for not following his previous instructions, saying: 'It appears that the Federal Defendants have abandoned the Conceptual Framework they committed to during the early phases of this remand.' Despite the judge's opinion, at the time of writing in early 2008, it looks as though the dams will stay and the salmon will roam less than free (Redden, 2007: 3).

Lessons learned

In many senses, the story of the Snake River foretells the future of water resources management. As societies achieve new levels of development and begin to look to different priorities, they must review anew how and why we manage our water. The management of the Snake River embodies some of the most complex issues encountered in applying IWRM principles: Federal and state government laws, regulations, agency policies, national and international treaties (between Canada and the USA and between the USA and the Indian Tribes), large and powerful NGOs fighting established industrial and commercial interests, water-quality as well as water-quantity issues, navigation, irrigation, water supply, and conflicts among rational economic calculations and subjective qualitative biological assessment. All of this is to be accomplished against a background of changing demographics, changing cultural values and new interests.

The title of this chapter is provocative, and it was meant to be. It highlights the conflicts between human needs and ecosystem needs. At one level, the only moral choice offered here is to pull down the dams and 'let the salmon roam free'. In theory a comprehensive, unified IWRM approach to problem solving is desirable (everything is connected to everything else); but in practice such an IWRM approach needs to go beyond simply being descriptive or a way of thinking. Instead, the approach needs to include prescriptive procedures for deciding and acting. Some intractable disputes will have to be resolved through legal mechanisms, which must be designed to reinforce and support good practice. Still, there also must be more give and take among legislative regulations and on-the-ground economic activities. It is not clear, however, how that increased give and take can be achieved under a regulatory approach rather than a planning approach. Regulations by themselves set up conflicting relationships among the parties rather than facilitating negotiations and tradeoffs among them. Given the litigious nature of the USA and the rigidity of the regulations, there is little hope for harmonious resolution of the perceptions of public interest.

In the case of managing any river basin in the USA, however, there are numerous external constraints – Federal legislation, state laws and regulatory administrative procedures which provide numerous avenues for veto power by well-organized interest groups enabled by legislation like ESA (Stakhiv, 2007). In this context IWRM needs to go beyond being a suitable vehicle for fact-finding, assessment, and formulating and evaluating options. Binding decisions require an institution with the authority to challenge and cut through the accumulation of overlapping rules, procedures,

regulations and conventional practices. Even the International Joint Commission, which manages water disputes between Canada and the USA and operates under a long-standing bilateral agreement, has difficulties enforcing its decisions.

The dam-removal issue in the USA is a messy one, tangled up in the Endangered Species Act, the prohibitive costs of taking dams down, and the negative environmental impacts of such a decision. As it turns out, those costs can far outweigh the benefits, particularly when the costs of recapitalizing replacement energy sources are taken into account. Small, run-of-the-river dams are not a problem; the Corps of Engineers has taken many down since the 1980s. However, projects with 120 million cubic metres or more will typically entail major conflicts among users and uses. The issue of removing the dams from the lower Snake River is not yet resolved. If it happens, it will be a decision enforced by the courts relying on ESA, demonstrating very clearly the importance of an effective legal framework to help achieve a balance among competing priorities as these evolve over time.

References

Bonneville Power Administration (2007a) *The Costs of Breaching the Four Lower Snake River Dams' Factsheet, DOE-BP–3777*, March 2007, available from www.bpa.gov/corporate/pubs/fact_sheets/07fs/fs030207.pdf

Bonneville Power Administration (2007b) *Replacing Lower Snake River Dams Would Cost Northwest $413 Million to $565 Million Annually, BPA Analysis, DOE-BP–3796*, April 2007, available from www.bpa.gov/corporate/BPANews/Perspective/2007/Snake_River_Dams/Replacing_Lower_Snake_River_Dams.pdf

IEAB (Independent Economic Analysis Board) (2007) *Review of the SOS Revenue Stream Report*. Independent Economic Analysis Board, Task No. 118, 25 February 2007, executive summary and link to full report available from www.nwcouncil.org/library/ieab/ieab2007–1.htm

Meadows, R. (2004) 'Struggling Against the Current', *ZooGoer*, vol. 33, issue 1, available from http://

nationalzoo.si.edu/Publications/ZooGoer/2004/1/P acific_Salmon.cfm

NMFS (National Marine Fisheries Service) (1995a) *Endangered Species Act – Section 7 Consultation. Biological Opinion. Reinitiation of Consultation on 1994–1998 Operation of the Federal Columbia River Power System and Juvenile Transportation Program in 1995 and Future Years*, National Marine Fisheries Service, Northwest Region

NMFS (National Marine Fisheries Service) (1995b) *Basis for Flow Objectives for Operation of the Federal Columbia River Power System*, National Marine Fisheries Service and the National Oceanic and Atmospheric Organization

Redden, Hon. James A. (2007) Letter of 7 December 2007 to Counsel of Record in Nat'l Wildlife Fed'n v. Nat'l Marine Fisheries Serv., CV 01–640 RE, and American Rivers v. NOAA Fisheries, CV 04–00061 RE

SOS (Save Our Wild Salmon) (2005) *Revenue Stream: An Economic Analysis of the Costs and Benefits of Removing the Four Dams on the Lower Snake River*, available from www.americanrivers.org/site/DocServer/revenuestre am8.pdf?docID=5121

SRSRB (Snake River Salmon Recovery Board) (2006) *Summary: Snake River Salmon Recovery Plan for SE Washington*, available from http://ice.ucdavis.edu/ education/esp179/?q=node/176

Stakhiv, G. (2007) US Army Corps of Engineers. Personal communication, August 2007

US Army Corps of Engineers (2002) *Summary: Improving Salmon Passage* in Final Lower Snake River Juvenile Salmon Migration Feasibility Report/Environmental Impact Statement, February 2002, available from www.nww.usace.army.mil/final_fseis/study_kit/study page.htm

Supplementary references

American Rivers. *Laws, Treaties, and Snake River Salmon Recovery*, available from www.americanrivers.org/site/ DocServer/Snake_Legal_factsheet.pdf?docID=1141

American Rivers. *American Rivers Fact Sheet: New $6 Billion Federal Plan for Columbia and Snake Rivers Allows Further Salmon Declines, Abandons Recovery*, available from www. americanrivers.org/site/DocServer/Final_BiOp_fact sheet_3–05.pdf?docID=1121

Anderson, J.J. *History of the Flow Survival Relationship and Flow Augmentation Policy in the Columbia River Basin*, working paper, June 2001, Columbia Basin Research, School of

Aquatic and Fishery Sciences, University of Washington, Seattle, WA, available from www.cbr.washington.edu/ papers/jim/flow_survival_history.html

Bonneville Power Administration. *The Snake River Dam Study: Then and Now*, factsheet, DOE/BP–3762. November 2006, available from www.bpa.gov/ corporate/pubs/fact_sheets/06fs/fs113006.pdf

Bonneville Power Administration. *Myths and Facts About the Lower Snake River Dams*, factsheet, DOE/BP–3763, November 2006, available from www.bpa.gov/ corporate/pubs/fact_sheets/06fs/fs113006c.pdf

Bonneville Power Administration. *Snake River Dam Removal*, 21 November 2006, available from www.bpa.gov/corporate/BPANews/Perspective/200 6/4_Lower_Snake_Dams/Snake_River_Dams.pdf

Federal Caucus News. *Federal Executives Make Statement About the Four Lower Snake River Dams*, 6 October 2006, PR 84 06, available from www.nwriverpartners.org/ documents/FederalExecutivesStatementabour4Snake RiverDams–10-6–06.pdf

International Rivers Network (IRN). *Reviving the World's Rivers: Dam Removal – Part 2: The US Experience*, 8 October 2007. Links to all six parts are available from http://internationalrivers.org/en/way-forward/river-revival/reviving-worlds-rivers

Milstein, M. *The Oregonian: Judge Rips Latest Plan to Help Salmon: Meeting with Judge James Redden is Federal Managers' Last Chance*, available from www.wildsalmon.org/ pressroom/press-detail.cgm?docid=729

Northwest Power and Conservation Council. *Analysis of Snake River Dam Removal has Deficiencies, Economists Report* press release, 14 March 2007, available from www.nwcouncil.org/library/releases/2007/0314.htm

Taxpayer.net. *The Salmon Economic Analysis and Planning Act*, available from www.taxpayer.net/snake/ seapafactsheet.pdf

Taxpayer.net. *GAO Shows Salmon Recovery Efforts an Expensive Failure*, available from www.taxpayer.net. To view the GAO report, go to www.gao.gov/new.items/ d02612.pdf

Taxpayer.net. *Members of Congress Urge Federal Government to Base Salmon Recovery Plan on Sound Science and Economics*, 22 June 2006, available from www.taxpayer.net/ snake/2006–0622BlumenauerPetriLetterRelease.pdf

Wikipedia. *Endangered Species Act*, available from http:// en.wikipedia.org/wiki/Endangered_Species_Act

Wikipedia. *Snail Darter Controversy*, available from http:// en.wikipedia.org/wiki/Snail_darter_controversy

6

Better Rural Livelihoods through Improved Irrigation Management: Office du Niger (Mali)

Boucabar Barry, Regassa E. Namara and Akiça Bahri

In many parts of the world, improving rural livelihoods and incomes means transforming the way in which agricultural water is managed. Mali is a case in point. The Office du Niger, formed in the 1930s as a centralized public enterprise to produce irrigated cotton and rice, has been significantly revamped since the 1990s. The result has been dramatic gains in rice production and farm incomes, and reductions in rural poverty. The case shows that changing agricultural water management requires a supportive macro-policy environment, and appropriate institutional changes and infrastructural investments. Equally important, it shows that those reforms may need to precede improvements in water management. Moreover, in aid-dependent low-income countries reform cannot occur unless both government and donors concur on the need for change. Finally, the case drives home that improving water management is a continuing process; gains to date in economic efficiency and (to a lesser extent) equity in the Office du Niger now need to be matched by improvements in environmental sustainability.

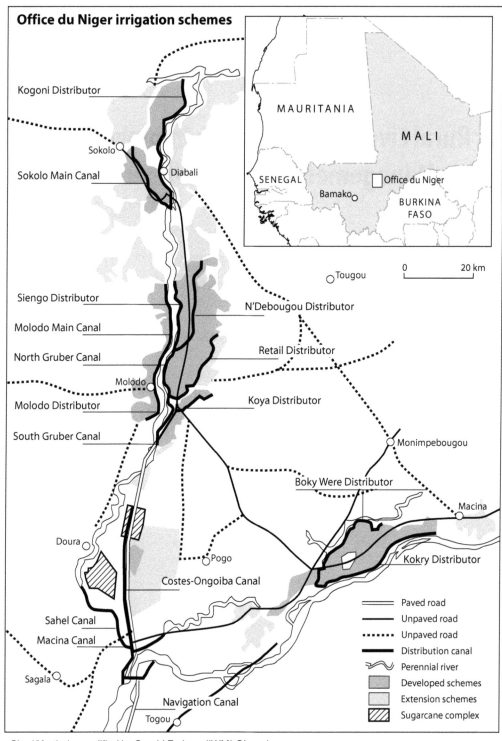

Source: Cirad/Karthala, modified by Gerald Forkuor (IWMI-Ghana)

Figure 6.1   The irrigated schemes of the Office du Niger

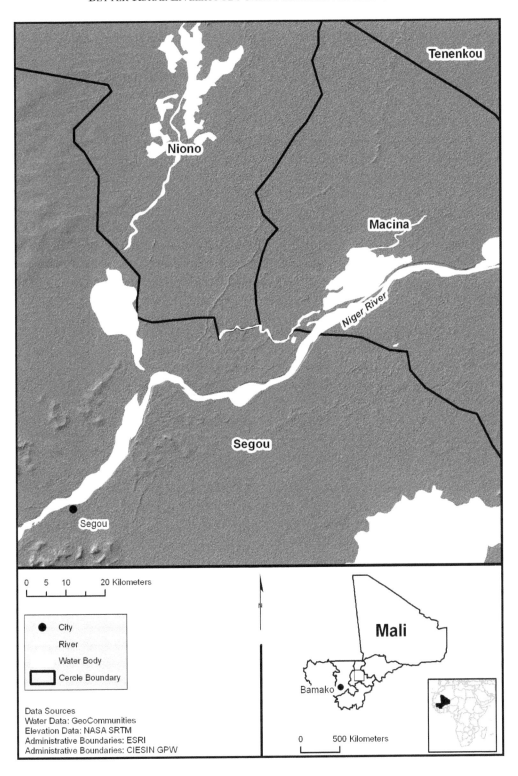

Location map

Water, agriculture and development in Mali

In 1925, a vast fossilized inland delta composed of alluvial soils well suited for gravity irrigation was identified between Segou and Timbuktu. Soon after, French colonial authorities established the Office du Niger in 1932 as an autonomous public enterprise for cotton-fibre production to reduce the French textile industry's dependence on imports, and to grow rice for French colonies in the Sahelian and Saharan regions. The plan was to develop the entire delta's 1 million hectare irrigation potential over 50 years. But 70 years later, just 7 per cent of that potential had been developed (Aw and Diemer, 2005).

The Office du Niger refers to both the 77,440 hectare irrigation scheme (see Figure 6.1) and the organization that controls it. It is in the heart of Mali, on the left bank of the Niger River. Rainfall in the Office du Niger ranges from 450 to 600 mm per year. Given Mali's arid to semi-arid climate, farmers and the 50 per cent of the country's GDP that depends on agriculture are extremely vulnerable to drought. The importance of the Office du Niger was highlighted by the major droughts in 1972–1974 and 1983–1984.

The Office du Niger had complex problems from the outset and its performance was disappointing. Bankruptcy nearly led to its complete dissolution in the late 1970s. In the 1980s a third of the developed area was abandoned as poor drainage and maintenance degraded the infrastructure and invasive weeds and wild rice spread. Average paddy yield declined from 2.5 to 1.5 tonnes per hectare (rice having replaced cotton production at the beginning of the 1970s).

The causes for the scheme's decay and falling productivity included inefficient water control, absence of production incentives for farmers, a very expensive bureaucracy and heavy reliance on state subsidies. Farmers were forced to sell their paddy to the Office du Niger, which enjoyed a monopoly on marketing and milling. Farm prices were low. Many farmers were unable or unwilling to pay the water fee. Fee income fell short of what was needed to maintain and operate the canals. In addition, numerous staff were engaged in rent-seeking.

In the early years, the Office du Niger controlled nearly all the goods and services within its realm; it was responsible for everything from surveying and developing land to building and operating health clinics and running literacy courses. This vertical integration was useful when the scheme was being built, because the Niger's interior delta was nearly uninhabited. Once roads and towns were built, however, vertical integration became a source of inefficiency and lack of accountability. Not surprisingly, during this period there was no involvement of stakeholders, little consideration of users' multiple needs for water, no attention to sustainable management and no recognition that water is a scarce resource.

At the beginning of 1990s, after many years of hesitation and negotiation, a set of donors (including the World Bank and the Dutch and French overseas aid agencies), and eventually the Government of Mali itself, realized the need for a profound and comprehensive technical, economic and institutional reform – to give farmers true ownership over the irrigation development process, and the capacity to sustain and manage irrigation infrastructure and operational and maintenance systems.

The approach to reform in the Office du Niger

The reform practices in the Office du Niger, whether by design or by accident, adopted some of the key principles of the integrated water resources management approach. The instruments used included measures to improve efficiency in water supply and use, a set of institutional changes, innovations in agricultural technology, the use of economic instruments and some important financing measures.

The reforms began with an analysis of the Office du Niger's problems, which looked at the interplay

Box 6.1   Key milestones

1932:	Office du Niger is founded with a focus on cotton and rice production.
1945–1948:	Abolishment of the forced recruitment of labour. Over 40 per cent of the settlers leave.
1960:	Mali gains Independence; nationalization of the land; around 55,000 ha developed.
1966:	Start of sugarcane plantation.
1970:	Cotton cultivation abandoned and conversion of cotton fields to rice.
1970–1976:	Land on state farms is redistributed to settlers.
1972–1974:	Severe drought throughout the Sahelian zone.
1978:	First donor meeting on Office du Niger rehabilitation.
1980s:	Reforms begin with investments to upgrade irrigation infrastructure.
1989:	Governance decree grants cultivation rights to farmers.
1990s:	Reforms broaden to the political arena and trade. Paddy price fully liberalized.
1991–1992:	Breakup of the Office du Niger's threshing and hulling monopoly.
1994:	Privatization or transfer of all economic activities to the farmers.
1994–1998:	Progressive reduction of import taxes on rice.
1995:	First performance contract between the Office du Niger, the state and the farmers. Redefinition of Office du Niger missions and responsibilities.
1995–1996:	Reorganization of credit.
1996:	Enhanced land tenure security for farmers. Farmers free to select threshers, mills, buyers.
1996–2003:	Diversification of crop production; increased cropping intensities and income.
1996–1998:	Second performance contract between the Office du Niger, the State and the farmers.
1997:	Creation of the first farmer trade union.
1997–1998:	Allotment of agricultural land use permits.
1999:	First elections of the rural communes councils.
1999–2001:	The level of water fees is set up; fee collection rate around 97 per cent.
2001–2020:	Preparation of master plan.

Source: after Couture et al, 2002; Aw and Diemer, 2005

between infrastructure, markets and institutions. By the 1990s, almost all of the Office du Niger's stakeholders including the Government of Mali agreed that its problems were institutional and structural. But donors and Government of Mali disagreed about how to address them. The Office du Niger favoured hardware and expansion; donors pressed for institutional reforms. Even the donors – the World Bank, the Netherlands (through the ARPON project[1]) and France (through the Retail project[2]) – differed on how to proceed; as a result each programme followed its own agenda in different parts of the scheme.

Improving water supply and use efficiency

Water supply and use efficiency was improved through rehabilitation and modernization, as well as by intensifying farming.

Physical rehabilitation/modernization

The donors gave priority to rehabilitation over expansion to improve the efficiency of water supply and use within the already developed irrigated lands. The Office du Niger's motto was expansion to hit the 1 million hectare target. The average yield was about 1 tonne of paddy per hectare, indicating the low efficiency of the use of water and land resources. Donors therefore urged physical rehabilitation activities to improve water transport and control using different formats, mirroring their own development experience and philosophy (Aw and Diemer, 2005). The cost of these physical modernization projects was substantial (Couture et al, 2002) – absorbing a major part of the external financing.

Farming systems intensification

The rehabilitation/modernization drive was complemented by new intensified farming including high-yielding rice varieties, transplanting, fertilizers, double cropping, crop diversification (mainly horticultural crops), better on-farm water management,[3] and resizing of fields into smaller plots to match the capacity of family workers. As with infrastructure, France's approach differed from that of the Netherlands. France prescribed cultivation practices for immediate adoption; the Netherlands took a medium-term view and relied on example, extension, farmer good sense and farmer experimentation.

Institutional triggers

The donors recognized the importance of a suitable environment to support irrigated agriculture and the need to emancipate farmers from the Office du Niger's excessive control. Their main priority was rightsizing the Office du Niger to reduce the Government of Mali's expenditures. For the most part they followed an indirect and gradual approach rather than a confrontational one. The institutional innovations that contributed to the eventual complete reform of the Office du Niger in 1994 focused on land tenure, support services, the role of the private sector, farmer organizations and representation, performance contracts and co-management arrangements.

Improving land tenure security

Land tenure was an obvious problem with no obvious solution. The Retail project made a lasting contribution to land tenure security – a farming licence procedure offering farmers two types of land tenure: permanent and temporary. Permanent tenure gave farmers after two years of probation the right to farm the same plot for an undetermined period, on condition of paying the water fee and practising intensive agriculture. Farmers had little incentive to intensify as long as they were tenants who might have to relocate. The new land tenure system shielded them from arbitrary relocation and linked the water fee to the quality of water service. Retail's other key contribution was to adapt farm size to family labour needed for intensive rice cultivation. The land tenure institutional reform improved farmers' rights not only to their rice fields, but also to the land on which their homes and household gardens were located.

Creating pro-farmer support services

In line with its policy of satisfying farmers' needs and empowering them to stand up to the Office du Niger bureaucrats, the ARPON project widened its array of support services. It established several new subsidiaries including the Agricultural Training Center in 1980; a non-commercial public works construction unit carrying out canal reconstruction work in conjunction with Office du Niger; a farm machinery assembly unit that produced and maintained affordable threshers, dehullers and farm implements, including the land levelling and puddling planks pulled by draught animals; and a seed farm and processing unit. In addition, both Retail and ARPON provided research support.

Strengthening the role of the private sector

Donors exerted persistent pressure on the Office du Niger to relinquish tasks that could be carried out by private entities. At the same time donors raised a pertinent question: will private entrepreneurs automatically and immediately take over tasks that the Office du Niger abandons? The Government of Mali and donors found that the emergence of a new private sector required some nurturing and guidance, since new providers of services need expert help to identify demand and gear up financially, managerially, legally, technically and environmentally.

Adopting performance contracts

The Government of Mali took advantage of donor differences on strategy to resist reform and to obtain more financial support. Only after the donors formed a united front in 1988 were they able to negotiate the first performance contact, which set firm three-year goals for the Government of Mali and the Office du Niger. In those first three years the Office du Niger became a public enterprise with full financial and administrative autonomy, and it made significant strides in improving infrastructure management. In 1994, the Office du Niger was restructured and has since been responsible only for allocating land, maintaining the secondary infrastructure and carrying out extension work.

Establishing co-management committees

In each of the Office du Niger's five zones, two Joint Committees were formed: one for system maintenance and one for land management. Roughly equal numbers of farmers' representatives and officials sit on these committees. They decide how and where to spend zone maintenance funds, check the spending and supervise the implementation of maintenance plans. The Joint Land Management Committees decide on land attribution and resolve land use conflicts.

Establishing farmer organizations

Donors (the Netherlands in particular) sought to empower farmers by organizing them into a single Village Association (VA) and stressed giving direct benefits to farmers. Donors set up functional literacy programmes, improved extension and reinforced social infrastructures. They hoped to help farmers deal collectively with the Office du Niger's managers.

Assigning farmer representatives

Two representatives speak on behalf of the farmers to the Office du Niger and other bodies such as the Ministry of Agriculture, the Chamber of Agriculture and local administrations. They sit on the Office du Niger's central non-executive board, which annually decides on water fees, the Office du Niger's budget and policy priorities. Through this board the farmers' representatives also negotiate with donor projects on rehabilitation of the irrigation and drainage system and other interventions. Their influence here, however, is less marked than on water fees and budget issues.

Innovations in agricultural technology

The rehabilitation and modernization projects described above were implemented partly as a way of challenging the power balance between the Office du Niger and farmers by giving farmers more control over canal management. Two technological innovations that helped break the Office du Niger's monopoly also contributed significantly to its eventual complete reform: small threshers and dehullers.

Introducing small threshers

ARPON equipped all VAs with small threshers at the ratio of one machine for each 100 hectares. The distribution of threshers broke up the Office du Niger's monopoly on threshing and significantly reduced the cost. The threshers allowed VAs to earn cash, which made it possible for them to offer credit to producers and to invest in health centres, mosques and schools. And the Office du Niger has not been able to re-establish its monopoly (Aw and Diemer, 2005).

Introducing small dehullers

Small dehullers – already in common use in Asia – helped farmers in two ways. First, they were the first step in breaking up the Office du Niger's rice milling monopoly, which had prevented small farmers and traders from taking advantage of the

1986 liberalization of the paddy market. Second, they saved women from spending hours each week pounding paddy for the family meal.

ARPON introduced small dehullers to the scheme by donating 80 of them to women's groups. The machines were more efficient, and resulted in lower transport costs because they were closer to where the clients lived than the Office du Niger's large mills. Although the small dehullers produced a lower-quality product than the Office du Niger's mills, consumers did not care. Soon, a market in rice milling arose. Rich farmers, traders and well-run VAs acquired larger dehullers to compete with the Office du Niger. Their investments drove down milling costs from US$0.0313 per kilogram of paddy in 1989 to US$0.025 per kilogram of paddy in 1992. Decentralization and the creation of a mass constituency with a stake in milling made untenable another part of the Office du Niger's monopoly (Aw and Diemer, 2005).

Economic instruments

The two key economic instruments employed as part of the change process were macroeconomic reforms and cereal market reforms.

Macroeconomic reforms

The economic environment of the scheme and farmers had improved with sectoral-level reforms and also major macroeconomic reform effected as a result of financial and political pressure from Bretton Woods institutions. In particular, the 50 per cent devaluation of the Central African (CFA) franc in 1994 greatly increased returns to tradable goods, including rice. In addition to benefiting from the scheme's rehabilitation, farmers in the Office du Niger profited from the liberal price climate in the first two years after devaluation, and were able to more or less dictate the price to buyers (Touré et al, 1997).

Cereal market reforms

Controlling the Office du Niger's rice was at the heart of the Government of Mali's cereals policy.

Cultivators were legally bound to sell their paddy to the Office du Niger, under threat of eviction, at prices set by the Government of Mali. In 1981, a forum called 'Projet de Réforme des Marchés Céréaliers' outlined a strategy to restructure the cereals market to:

1 raise producer prices and thus stimulate more production;
2 liberalize the market for cereals to stabilize the supply; and
3 redefine the role of the Government of Mali to raise savings with which to upgrade civil servant pay and purchasing power.

In 1982, the Government of Mali officially abolished its monopoly on handling agricultural produce, but measures liberalizing the rice market were not taken until the end of the 1980s, and then only reluctantly. Liberalization allowed farmers to obtain substantially greater returns on their production, though some individual farmers lost large sums of money on contracts with dishonest traders.

Financial measures

Because they were deep in debt, many farmers were ineligible for official credit. The ARPON project managers argued that credit was necessary to break the downwards spiral of poor crop husbandry, low yields and an inability to pay water fees, which taken together usually ends with the farmer's eviction. In addition, inadequate resources to maintain irrigation infrastructure leads to an unreliable and inadequate water supply. ARPON provided seasonal credit for low doses of fertilizer (75 kg urea and 50 kg DAP per hectare), and five-year credit for draught animals and implements.

In addition, the Retail project devised a way to ensure adequate financing for infrastructure maintenance. It created a dedicated maintenance fund to which it allocated 70 per cent of the water fee proceeds. A joint committee – balancing Office du Niger and farmer representatives – controlled the fund. Office du Niger staff initially resisted this reduction of their powers. ARPON adopted the idea of a dedicated maintenance fund jointly managed by Office du Niger staff and farmer representatives. What began as decentralized village development funds run by the VAs later evolved into village-level savings and loans associations.

The outcomes: efficiency, equity and sustainability impacts

The infrastructural, institutional, economic and financial changes introduced in the mid-1990s have had very substantial impacts on economic efficiency and (to a lesser extent) equity in the Office du Niger. Progress on environmental sustainability, however, has been slow.

Efficiency gains in business process and increases in cultivated area

The Office du Niger's staff was dramatically reduced to 360 from nearly 4000, while its budget was reduced to 2–3 billion CFA from 6–10 billion CFA (to US$4–6 million from US$12–20 million). The water service fee declined from 20 per cent to 8 to 9 per cent of net farm income; the fee collection rate increased from 80 to 97 per cent. Reducing its salary burden has allowed the Office du Niger to allocate a much greater share of the revenues generated from irrigation fees to the operation and maintenance of the network, thereby contributing to sustainability. More important, user committees ensure that settlers are treated as full partners regarding the utilization of fees. It has also been possible to raise the fees by 43 per cent in order to fully recover operation and maintenance costs.

Before the reforms, the cultivated area had not expanded. Since 1996 the cultivated area has increased by more than 20 per cent through farmers' participation and donor and private investment. In the past mono-cropping of rice was

the norm, with only 75 per cent cropping intensity. Since the reforms the cropping intensity has climbed to 120 per cent. The area planted with horticultural crops has increased from 1929 hectares in 1996 to 5,245 hectares in 2003.

Improvements in rice productivity and production

In Mali, rice now accounts for 12.3 per cent of agricultural value added, up from only 4.3 per cent in the 1980s. Rice is the fastest growing agricultural sub-sector, averaging 9.3 per cent annual growth (Gagnon, 2005). Since more than half of Mali's total rice production comes from the Office du Niger, it is fair to attribute the phenomenal growth rate in Mali's rice production to the reforms (Kelly and Staatz, 2005). The production of rice was about 427,800 tonnes in 2002–2003, representing 62 per cent of the total 693,203 tonnes harvested in Mali and 57 per cent of the country's requirement (Figure 6.2).

The increased rice production is a result of increases in both acreage and yield. The total area under rice cultivation within the Office du Niger is estimated at 77,440 hectares, which accounts for about 21 per cent of the total paddy fields of the country (Figure 6.3). The increase in area is mainly through the reclamation of previously abandoned land (about 10,000 hectares), since no new land development has been undertaken for rice since 1966.

In 2008, following two decades of reform, average yields approximate 6 tonnes per hectare, with many farmers achieving yields of 7 and even 8 tonnes. Average paddy yields have tripled, which compares favourably with the Green Revolution achievements in Asia. The average yield in the Office du Niger has constantly increased over 22 years, from 1.6 tonnes per hectare in 1982–1983 to 6.1 tonnes per hectare in 2002–2003 (Figure 6.4).

Improved irrigation and rice production technologies, combined with market reforms that created incentives for farmers to adopt them, are at the heart of productivity growth in the Office du Niger. The new technologies made a difference

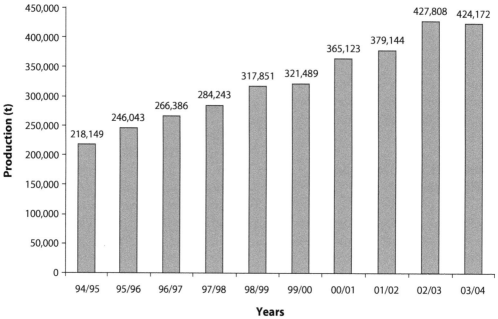

Source: Office du Niger records

Figure 6.2   Rice production trend at the Office du Niger, 1994–2004

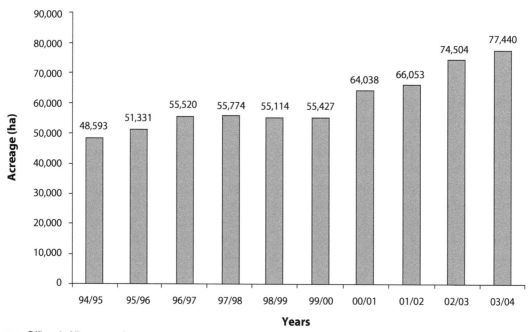

Source: Office du Niger records

Figure 6.3   Paddy fields area trend at the Office du Niger, 1994–2004

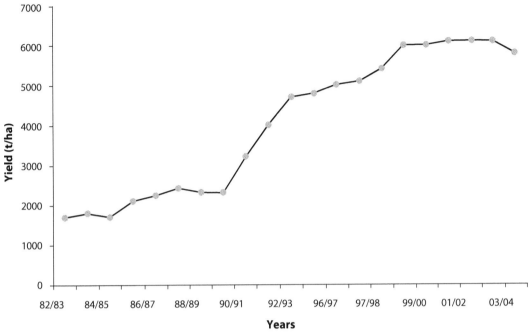

Source: Office du Niger records

Figure 6.4   Paddy yield trend at the Office du Niger, 1982–2004

because the economic and institutional environment evolved to provide a broad range of socioeconomic incentives for both farmers and service providers.

Improvements in employment and labour productivity

Before the mid-1980s, rice production in the Office du Niger was extensive, with a family of 3.5 active workers cultivating 5 to 6 hectares (about 1.7 hectares per adult worker). The rice system has become more labour intensive; labour shortages are increasingly common at peak periods when the rice is being transplanted or harvested. Young people coming from the Sahel are contracted to work for a task, a day or a season. Transplanting has become a specialized task done mainly by women. Attracted by a chance to earn money, the settler population has grown quickly. In 2004, there were 364,769 inhabitants living in 253 villages within the Office du Niger area, a 110 per cent increase since 1934–1947. Under conditions of natural growth,

without immigration, the population would have increased by only 27 per cent. Population growth was due in part to rising labour productivity: paddy production per person increased by 75 per cent (Figure 6.5). More expenditure on hired labour was associated with higher rice yields, a reflection of the very labour-intensive nature of rice production.

Population growth was spurred also by income opportunities arising from hardware modernization and new services in support (training centre, input fund, equipment assembly, seed farm, farming system research and development), processing and trading.

Differential changes in farm income

Crop budget analyses consistently show high levels of net returns per hectare for most Office du Niger farmers (Bélières and Bomans, 2001; Mariko et al, 2001; Chohin-Kuper et al, 2002) despite substantial increases in the cost of fertilizers and the water

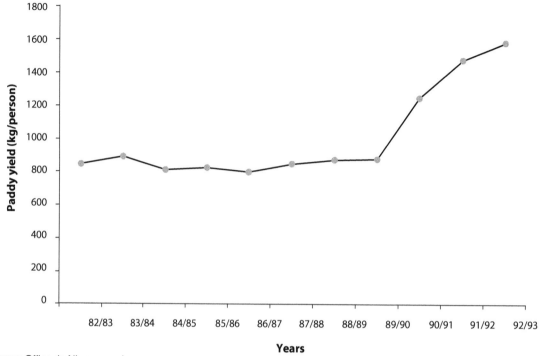

Source: Office du Niger records

Figure 6.5   Per capita paddy yield trend in the Office du Niger

users' fee. The Office du Niger's estimated average crop budget for the 2003 and 2004 rainy season rice production shows that the gross margin per hectare is almost 250,000 CFA (approximately US$500).

The one weak spot financially is those farms with fewer than 2.5 hectares of land, which cannot provide subsistence incomes for farming households. Table 6.1 provides a comparison of gross margins per person and per hectare calculated from the 1999 survey data of a sample representative of the entire Office du Niger. The differences across farm size are striking, particularly for the gross margins per capita. They suggest that small farms are probably having problems covering consumption needs after paying input costs. More recent studies (Kébé et al, 2003) show the share of the consumer price retained by farmers to vary between 70 and 90 per cent, depending on the time of year.

An increasingly important income source for farmers in the irrigated perimeter has been the

Table 6.1   *Comparison of gross margins for rice production by farm size*

Type of farm	Number of farms (FCFA)	Average cultivated area per farm (ha)	Gross margin per farm (FCFA)	Gross margin per ha (FCFA)	Gross margin per person
Small [0.25 to 3 ha]	254	1.86	482,382	259,345	59,208
Medium [3 to 10 ha]	260	5.61	1,587,977	283,062	105,686
Large [> 10 ha]	35	14.72	4,524,759	307,389	141,901
All farms	549	4.46	1,263,690	283,338	86,492

Source: Bélières and Bomans, 2001

Table 6.2 *The extent of vegetable production in Office du Niger (1995/1996–2003/2004)*

| | Shallot | | Tomato | | Sweet potato | | Pepper | Maize | Potato | | | |
	ha	tonne	ha	tonne	ha	tonne	ha	tonne	ha	tonne	ha	tonne
1995/1996	1,365	32,760	125	2,500	—	—	110	935	—	—	—	—
1996/1997	1,628	42,393	150	3,059	55	990	96	848	141	382	—	—
1997/1998	2,551	66,419	147	2,714	69	1,242	150	848	155	466	—	—
1998/1999	2,665	68,839	227	3,975	86	1,572	105	464	385	1,194	—	—
1999/2000	2,274	61,012	227	3,975	97	2,651	105	464	470	1,550	31	470
2000/2001	3,271	68,600	430	8,755	283	6,842	132	831	533	1,864	35	699,
2001/2002	2,588	72,462	387	8,162	517	11,097	164	878	607	2,264	41	808
2002/2003	3,854	107,260	499	9,886	459	10,944	175	910	1,064	3,845	129	2,778
2003/2004	3,861	123,721	345	7,376	471	10,163	264	1,063	952	3,779	134	2,859

production of horticulture for domestic and regional markets (Table 6.2).

Impacts on poverty

The increase in income has improved livelihoods. For example, the gross income generated during the 2003/2004 season was estimated to be 80 billion CFA (US$160 million), with 50 billion CFA (US$100 million) from rice and 30 billion CFA (US$60 million) from vegetable production. Per capita farm income increased by 150–200 per cent. This is significant poverty reduction, although there are distributional issues with incomes skewed in favour of larger land-holders. The impact on poverty is even higher when one takes into account that many people flocked into the area. Generally, Mali's food security was enhanced because Office du Niger production is not subject to variations in the climate. The country approached self-sufficiency in rice in normal years (in which no flood, drought, crop pest or disease outbreak is experienced). New activities such as transplanting, rice processing and rice marketing provided new revenues to women and youths.

Spin-off effects

The Office du Niger, and particularly the reform, has also indirectly contributed to income growth and poverty reduction through multiplier effects in sectors connected to irrigated agriculture. The non-farming population is mainly involved in transportation and post-harvest processing of agricultural products, marketing, and maintenance of the various infrastructures. For example, in 2002, the Office du Niger paid 1.8 billion CFA (approximately US$3.6 million) to about 150 entrepreneurs who maintain the network of primary and secondary canals, and 3 billion CFA to transporters (US$4 million for transporting the rice and US$2 million for the vegetable products). Small private businesses have set up threshers and mills into the area.

Environmental impacts

Inefficient use of irrigation water and problems with drainage, waterlogging and salinity have always plagued the Office du Niger. More recently, the spread of water-borne diseases, the wide use of fertilizers and pesticides, and increased deforestation have also caused concern. So far, however, the reforms have not paid much attention to these and other environmental problems.

One key problem is inefficient use of water. One cause of that inefficiency may be that the Niger River has abundant water relative to current irrigation needs. The total amount of water abstracted from the Niger River is less than 10 per cent of its flow from July to December. Abstractions are much more significant during the dry season cropping, although less area is cultivated. While there is clearly enough water to meet current needs, future expansion of irrigation in the Office du Niger will depend on improving water management practices, since at present only 25 per cent of the water from the Markala into the Fala de Molodo and the Retail Scheme is effectively used by crops (Ouvry and Marlet, 1999). In addition, 44 per cent of the water in the primary canal network is lost to evaporation, 21 per cent in the drainage network especially during the growing season, and about 10 per cent of the water is lost by infiltration from the canal mainly during the dry season cropping.

A related problem is poor drainage. The principal causes are that the drainage system is saturated with excess irrigation water and the collector drains are not well maintained. Once the collector drain is no longer saturated, maintenance of the tertiary drains becomes important, while the disparity in harvest dates between adjacent fields always aggravates the drainage problem. The drainage problem has led to waterlogging, a fast rise in the water table and land degradation due to secondary salinity. According to Marlet and N'Diaye (1998), before the rehabilitation schemes, management of irrigation and drainage was deficient, which increased the alkalinity of clay soils (sandy soils were not affected). After the rehabilitation phase, water management improved and more diverse crops were grown during the dry season. The pH and EC (electrical conductivity) of the sandy soils have been increasing because in most parts of the scheme the canals contain water year round, which raises the water table.

Water-borne diseases have also caused concern. A study on human health conducted by SOCEPI in 1998 concluded that rice cultivation in areas under the Office du Niger has increased the prevalence of water-borne diseases such as malaria, bilharziosis and diarrhoea. Wide use of fertilizers and pesticides has increased the risk of contamination of groundwater as well as surface water. Rapid population growth has resulted in more deforestation because of a high demand for wood for fuel.

Lessons learned

Managing an irrigation infrastructure that was planned and developed without due consideration to farmers' objectives and realities is prone to various inefficiencies. The Office du Niger's initial goal was to supply the French textile industry with cotton fibre; the farmers had little or no prior experience with irrigated agriculture. The long years of adjustment needed to bring the scheme to its current state therefore are unsurprising. Change came gradually through technical and political improvements as opportunities arose, or through negotiation and donor pressure (as with land tenure) when they did not.

Below we highlight some of the important lessons learned from the Office du Niger's long years of turbulence and eventual success. These lessons are organized in four categories: the impact of the macro-policy environment on the process of change; aligning the interests of government and donors; enhancing the role of farmers; and improving the balance among economic efficiency, equity and environmental sustainability

The impact of the macro-policy environment on the process of change

The right environment

Restructuring has a much better chance of success and sustainability when it occurs in the context of the right macro-policy and with the right technology package. The impressive gains from the investments made in the Office du Niger infrastructure and the privatization of marketing and processing activities created momentum and support for the Office du Niger's restructuring. Macro-economic reforms translated into better producer prices, which provided both the incentive and the means for better use of all resources including water.

Importance of the coherence and timing of the reforms

The investments to upgrade the irrigation system, the liberalization and the timing of the devaluation and import policy reforms all helped the reform process, since they made Malian rice competitive with Asian imports in Mali and in neighbouring countries. Importantly, the process of reform in the Office du Niger has proceeded in stages: from hydraulic and agronomic targets and outcomes to economic and financial targets and outcomes. Note that financial targets have not yet been reached because of low water fees (Couture et al, 2002).

The role of government and donors – aligning interests

Government commitment and the participative approach

Overhauling the Office du Niger required the commitment not only of the Ministry of Rural Development, but also of the entire government, and particularly of the Ministry of Finance. The Government of Mali firmly committed itself to restructuring the Office du Niger by setting up an ad hoc unit directly under the authority of the Prime Minister; its sole mandate was to carry out the restructuring in three years. Building consensus and commitment in a participative way around the key objectives and principles of restructuring takes time, but it is vital to success.

Donor coordination

In aid-dependent low-income countries such as Mali, making key policy changes and restructuring inevitably requires strong donor coordination, especially when the changes are politically or socially delicate. In the case of the Office du Niger project, joint donor missions have contributed to building a strong common platform. Also, in 1992, coordination and team work with the IMF resulted in a very unambiguous joint message to the government regarding the urgency of restructuring.

Enhancing the role of farmers

Focus on water management with transparent financing

Refocusing the irrigation institutions on the essential water services with well-defined sources of funding is key to success and sustainability. The Office du Niger funds water maintenance entirely from producers' water fees.

Land tenure

Farmers need to feel land secure before they will invest in their fields. The Office du Niger's experience with usufruct rights shows that farmers can feel land secure without actual landed property. In several areas farmers are now willing to finance the irrigation investments at the field level.

Partnership with producers

Building a partnership with producers is essential to successful and sustainable irrigation schemes. The partnership between farmers and the Office du Niger was a dramatic departure from the top-down approach of the past; it has already generated high payoffs, such as the sharp increase in water fees recovery.

Improving the balance among economic efficiency, equity and environmental sustainability

Despite the Office du Niger's significant advances much remains to be done. Some further gains in economic efficiency could be achieved by further integrating horticultural production into the system, integrating and modernizing the livestock sector, and better managing natural resources. Expanding the irrigated area could also further increase economic gains, though it will also eat into the neighbouring grazing lands and affect natural resource management in the Office du Niger. If the number of livestock continues to grow, pressure on the remaining grazing lands will increase. An environmental study by the Office du Niger concluded that expansion must include measures to reduce environmental impacts within and outside the irrigated area. If major problems are to be averted, livestock production will have to intensify and animal movements within the irrigated area be controlled.

From the discussion above, it is clear that better water management was only one of several features of the Office du Niger reforms – and not the leading one at that. Given the availability of the water resource and the importance of the other constraints, however, arguably it was appropriate to begin with other reforms, which have already led to improved water management and have created a framework for future improvements. All this highlights the importance of sequencing and shows that improvements in the management of the water resource may not necessarily be the first step in a process of change.

A further broadening of the approach taken thus far is needed to further improve the Office du Niger's efficiency, equity and especially its environmental sustainability. The gains in economic efficiency and (to a lesser extent) equity in the Office du Niger are impressive. Now more attention to environmental sustainability is warranted. Such improvements would help to improve the physical and economic productivity of water. More rigorous water management is required because of the expansion of the irrigated area and the growing diversification of cropping systems. Enhancing water efficiency and improving drainage, together with increasing intensification, would increase the Office du Niger's overall performance. These measures would also offer maximum protection to the wetlands located on the eastern side of the inner delta. Changes in canal and field water management are also needed to protect the health of some 170,000 residents who still suffer losses in quality of life and productive capacity from water-borne diseases. Government investment in land development needs to be weighed against the impact of public health investments on productive

capacity. Further progress will require looking beyond water for crop production and considering all of a community's other water needs such as for domestic use and sanitation, livestock and fisheries, and ecosystems services.

As the Office du Niger's experience shows, improving water resources management is inevitably a continuing process – it simply never ends. The emphasis on improving economic efficiency in the Office du Niger now needs to be matched by attention to environmental sustainability.

Notes

1. Amélioration de la Riziculture Paysanne à l'Office du Niger (Improvement of Peasant Rice Cultivation at the Office du Niger).
2. A primary canal on which France sponsored land rehabilitation, crop intensification and institutional reform projects.
3. Better water management interventions include controlling water levels and flows with new sluices; and coordinating and redefining the roles of lock keepers, water inspection agents, arroseurs chiefs, etc.

References

Aw, D. and Diemer, G. (2005) *Making a large scale irrigation scheme work: A case study from Mali. Directions in development*, Washington, DC, The World Bank

Bélières, J.-F. and Bomans, E. (2001) *Coût de production du riz de contre-saison et d'hivernage 1999 dans la zone Office du Niger. Résultats partiels des enquêtes détaillées sur les exploitations agricoles de la zone Office du Niger*, Note No. 2, Ségou, Mali

Chohin-Kuper, A., David-Benz, H. and Mariko, D. (2002) 'La rémunération des riziculteurs. Une amélioration de la compétitivité du riz local au profit des producteurs, in Bonneval, P., Kuper, M. and Tonneau, J.-P. (eds) (2002) *L'Office du Niger, grenier à riz du Mali*, Paris, CIRAD/Karthala

Couture, J.-L., Delville, P.L., Spinat, J.B. and Cluster, E.D.R. (2002) 'Institutional innovations and water management in Office du Niger (1910–1999): the long failure and new success of a big irrigation scheme', *Coopérer aujourd'hui*, No. 29, pp1–57

Gagnon, G. (2005) *Mali Country Economic Memorandum (CEM) Agriculture and Rural Development*, consulting report for the World Bank

Kébé, D., Coulibaly, B.S., Traore, A. and Dembele, B. (2003) *Agricultural Growth and Poverty Reduction in Mali*. Paper prepared for the Roles of Agriculture International Conference 20–22 October 2003, Agricultural and Development Economics Division (ESA) of the Food and Agriculture Organization of the United Nations, Rome

Kelly, V. and Staatz, J. (2005) *Mali: Country Economic Memorandum, Agricultural and Rural Development*, Department of Agricultural Economics, Michigan State University

Mariko, D., Chohin-Kuper, A. and Kelly, V. (2001) 'Libéralisation et dévaluation du franc CFA: la relance de la filière riz à l'Office du Niger au Mali', *Cahiers Agricultures*, vol. 10, pp173–184

Molden, D., et al (2007), 'Pathways for increasing Agricultural Water Productivity', in Molden, D. (ed.), *Water for food, Water for Life: A Comprehensive Water Assessment of Water Management for Agriculture* Earthscan and IWMI, London and Colombo

Touré, A., Zanen, S. and Koné, N. (1997) *La restructuration de l'Office du Niger. Contribution d'Arpon 3 (Coopération Néerlandaise)*, Office du Niger, Ségou

Ouvry, F. and Marlet, S. (1999) *Suivi de l'irrigation et du drainage, étude des règles de gestion de l'eau et bilans hydrosalins à l'Office du Niger (cas de la zone de Niono, Mali)*, PSI-Mali, étude et travaux No. 8, IER, Bamako: tome 1, synthèse des travaux, 30pp; tome 2, compte-rendu d'expérimentation, 105pp + annexes

Marlet, S. and N'Diaye, M.K. (1998) *Evolution temporelle et variabilité spatiale des indicateurs de la dégradation des sols par alcalinisation et sodisation à l'Office du Niger*, PSI-Mali, étude et travaux No. 6, IER, Bamako: tome 1, synthèse des travaux, 53pp; tome 2, compte-rendu d'expérimentation, 124pp + annexes

7

From Water to Wine: Maximizing the Productivity of Water Use in Agriculture while Ensuring Sustainability

Simi Kamal and Amina Siddiqui

A combination of drought in key food producing areas of the world and growing demands from the expanding economies in Asia and elsewhere has drawn attention to the challenge of providing food for the world's growing population. It is increasingly recognized that one challenge of producing food is finding the water to do so.

An example of the challenge comes from the arid continent of Australia, whose farmers' exuberant use of water for agriculture has become increasingly constrained. This has been felt particularly acutely in the Murray–Darling river system, which underpins much of the country's rural economy.

The experience of a small community of farmers on the Angas Bremer irrigation scheme, at the mouth of the Murray–Darling river system in South Australia, provides an example of how better water management can contribute to meet emerging challenges. Rigorous control of water use combined with changing patterns of agricultural production, backed by an effective river basin organization, national policy and interstate agreements in a federal system, have all enabled the farming community to adapt to the pressures and even improve environmental conditions along with their productivity.

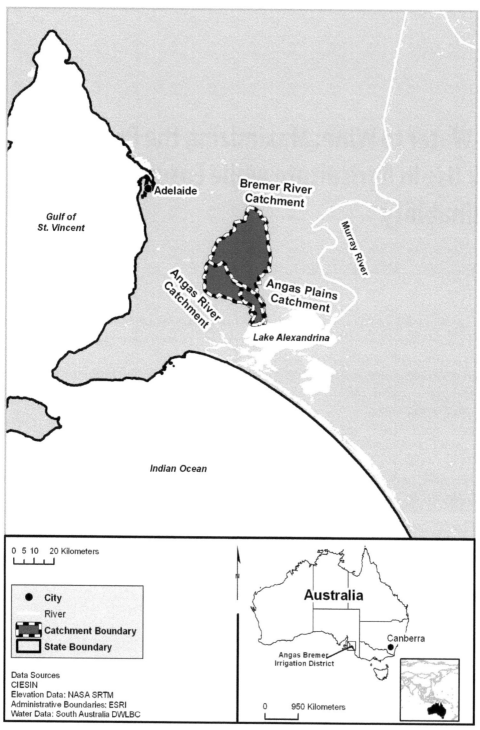

Source: Angas Bremer Water Management Committee Inc. (www.angasbremerwater.org.au)

Location map

The development context and water challenges

The Angas Bremer Irrigation District (ABID) covers a community of some 160 irrigators who farm an area of about 6,800 hectares on the Angas Bremer floodplain, near Lake Alexandrina in the Langhorne Creek area, 30 kilometres from the mouth of the Murray River.

It is an area rich in alluvial soils at the terminus of the relatively small Angas and Bremer river catchments where, historically, settlers were attracted by the fertile soils and the (then) abundant freshwater in the coastal lakes and the rivers that fed them. The farming community drew on these resources to support a variety of crops, notably lucerne hay for use as stock feed.

In the early years of settlement, in the 1860s, there was no need for formal irrigation. River flow from winter rains flooded the riverine plains and provided enough water for the growing season. Later, in 1886, some farmers built a weir to create artificial flooding in dry years. Other farmers opted for their own systems, including steam-driven pumps and windmills to lift water from the rivers and wells to their lands (ABWMC, 2005).

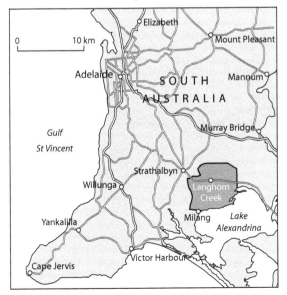

Figure 7.1   The Angas Bremer irrigation district

As agriculture developed further from the rivers, other sources of water were required. From the 1950s when state-provided electricity became generally available in the area, farmers began to pump groundwater, which appeared to be in plentiful supply.

Groundwater development was accelerated when the quality of surface water began to deteriorate in some areas due to a combination of lower flows in the rivers and pollution from mining and other activities upstream.

During the 1960s and 1970s there were unchecked withdrawals of groundwater in the Angas Bremer region. By 1981 annual usage of groundwater was four times the annual rate of recharge from flows down the Angas and Bremer Rivers. This excessive use led to declining quality of the groundwater, as saltier groundwater was drawn in from the margins. This in turn obliged the farmers to turn to the nearby Murray River for sustainable supplies.

The Murray–Darling Basin

The Murray–Darling Basin, which covers the catchments of the Murray and Darling Rivers, occupies an area of over 1 million square kilometres and accounts for about three-quarters of the total area of irrigated crops and pastures in Australia. Around 70 per cent of all water used for agriculture in Australia is used for irrigation in the basin. The over 2 million people who live in the basin and another million outside the basin rely on its water resources.

As the state at the 'tail end' of the Murray–Darling river system (of which the Angas and Bremer Rivers form a part through their interconnection with the coastal lakes), South Australia had long suffered the impact of upstream water uses as well as from perennial droughts.

The experience of the farmers in the Angas Bremer Irrigation District, and indeed of South Australian irrigated agriculture as a whole, demonstrated the need for a comprehensive management approach to the whole Murray–Darling Basin on which so much of Australia's agriculture depended.

A long drought – the 1895–1903 'Federation Drought' – drove a process of cooperation that eventually resulted in the 1915 River Murray Waters Agreement between the governments of Australia, New South Wales, Victoria and South Australia. It took two more years to establish the River Murray Commission, which was given the responsibility of putting the Agreement into effect. The division of water negotiated between the states at that time and formalized in the Agreement is effectively still in place (see Figure 7.2).

The entitlement flow for South Australia under the Murray–Darling Basin Agreement determines the minimum flows that South Australia will receive across the Victoria border. The entitlement flow is made up of the so-called 'Cap', the maximum allowable use (see page 97) with the balance being available for environmental use, system losses (evaporation and seepage) and system maintenance. During periods of low flow, these figures are adjusted by the formal processes outlined in the Murray–Darling Basin Agreement, which also provides drought arrangements.

The approach, triggers and processes

If it was the Federation Drought that triggered the first interstate agreement on the Murray in 1915, it was the more local threat of losing much of their irrigable land to salinization that saw the farmers of the Angas Bremer Irrigation District coming together in the Angas Bremer Water Resources Committee (ABWRC), a statutory body of local irrigators and government agencies, which was formed in 1978. The (South Australian) Water

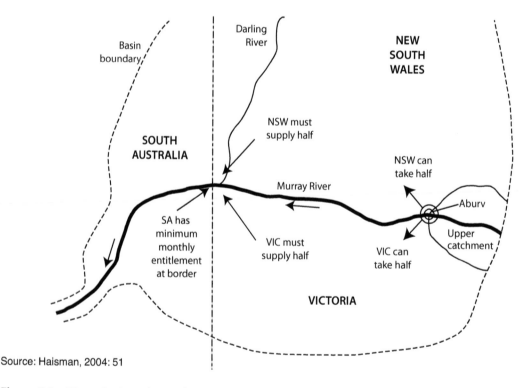

Source: Haisman, 2004: 51

Figure 7.2 Water sharing rules on the Murray River

Components within SA's Entitlement Flow

- • SA water (130 GL average per year)
- • Irrigation – Lower Murray (103.5 GL)
- • County towns (50 GL)
- • Other consumptive uses (440.6 GL)
- • Other – environment/maintenance flows

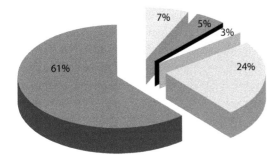

Source: SAMBNRMB, 2007: 11

Figure 7.3 Components within South Australia's entitlement flow

Resources Act of 1976, which is credited with being Australia's first integrated water resources management legislation (Schonfeldt, 2000), provided the institutional framework for these water stakeholders to work together.

One of the first key challenges of the ABWRC was managing the groundwater resource, which had been developed beyond its sustainable limits. After some years of monitoring, it had become clear that the amount of water allocated was perhaps four times more than could be sustainably abstracted, given the rate of recharge from the Angas and Bremer Rivers. Yet there was growing pressure for further development, driven according to one author, Muller (2002), by significant rises in the price of grapes. In 1981, under the jurisdiction of the ABWRC, the Angus Bremer Prescribed Wells Area (ABPWA) was proclaimed.

Between 1981 and 1987, attempts to resolve the problem of increasing groundwater salinity by reducing water allocations and encouraging aquifer recharge were unsuccessful. Faced with a growing

crisis, the South Australian Minister of Water Resources decided in 1988 that a substantial proportion of licences for underground water should be exchanged for water direct from the Murray River.

This development was driven in part by demand, as it became clear that a new farming mix – focusing on grapes and vegetables – could support water supply costs that the old mix could not. And, during this process, the irrigated area grew from 3,400 hectares (mainly under lucerne hay) in 1981 to 6,800 hectares in 2001, of which only 471 was under lucerne hay while the bulk came under vines.

Structure and functions of the Murray–Darling Basin Organization

The strategies adopted in the Angas Bremer linked its future firmly with that of the Murray–Darling Basin. The reforms that had led to the establishment of the Murray–Darling Basin Commission (MDBC) reflected a series of larger challenges, many of them environmental. A World Bank research paper commented that:

> There is presently a major focus of debate in Australia concerning the need to recover a proportion of water now allocated to agriculture and to re-assign it to the maintenance of river health. Assuming the resumption of property rights from agriculture will be accompanied by compensation payments, the total cost could be in excess of $2 billion (approx. US$ 800 million).
>
> (Haisman, 2004)

At the same time, public administrators were becoming ever more cognizant of the fact that natural resources tended to interact one with another and that a catchment or river basin made a sensible administrative unit within which to attempt to manage these processes. This was accompanied by a slowly growing awareness that natural resources were not infinite and that use of resources had tangible consequences that frequently offset project benefits.

'The impetus for this awareness was a combination of scientific research and evaluation of the status of natural resources valued by the community, lobbying by conservation interests, and the occasional crisis that excited the attention of popular news media. Two such crises relevant to this paper were the closure of the mouth of the Murray River because of drought conditions in 1981, and the "world's longest algal bloom" extending over 1000 kilometres of the Darling River in 1990–91. Although the highly variable Murray River could naturally cease to flow in its upper reaches, and indeed the explorer who found the Murray mouth recorded having to drag his boats over sand bars to reach the ocean, the idea that the nation's mightiest river could fail to reach the sea and that water extractions might have something to do with this, struck a responsive chord in distant urban populations' (Haisman, 2004: 16).

The establishment of the MDBC as the executive arm of the Murray–Darling Ministerial Council (MDMC) reflected these trends with its focus on integrated catchment management. The figure below shows the functions of the MDBC in terms of natural resource management and Murray River water.

South Australian Murray–Darling Basin strategy

While there was now a stronger institutional framework at the level of the river basin, much water management continued to be undertaken at the level of the state and depended on state institutions. In South Australia, the state government prepares a State Water Plan to set the strategic directions for water resources management. Individual catchment water management boards (covering sub-catchments of the overall basin) are then required to prepare catchment water management plans to establish, and provide for the implementation of, catchment priorities, as well as to prepare water allocation plans where required. These have to ensure that environmental flow requirements are met before water is allocated for commercial use.

The water reallocations that were made in the Angas Bremer area would have been hard to implement without the national-, basin- and state-level strategies and catchment-level boards. And the South Australian River Murray Catchment Board (RMCB) became the key interface between the Angas Bremer irrigators and the multitude of other users in the Murray–Darling system.

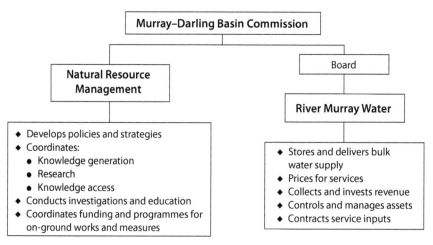

Source: Haisman, 2004: 58

Figure 7.4   Internal functions of the Murray–Darling Basin Commission

But reallocation was just a first step as the Angas Bremer moved further down the road towards economic and environmental sustainability. If the larger objective of reallocation from existing water users to the environment was to be met, all users, including those in the Angas Bremer would have to be engaged.

Instruments used

Over a period of nearly 50 years, an array of regulatory and technical instruments have enabled the irrigators of the Angas Bremer not merely to sustain their operations, but to expand their scope and profitability while at the same time improving the condition of the water resource.

Groundwater management plans

The initial innovation was to introduce and enforce local groundwater management. About 85 per cent of the groundwater in the Angas Bremer Prescribed Wells Area (PWA) comes from two low-salinity zones located along the Angas and Bremer Rivers. One of the earliest steps under the Angas Bremer Water Allocation Plan was to implement licensing as a means to control depleting groundwater. Structured efforts were also made to recharge the aquifers in good rainfall years, which helped to reduce depletion. An Aquifer Storage and Recovery system was initiated and promoted in the area in the 1980s, through which irrigators directed winter floodwaters into their bores.

A groundwater monitoring network had already been established in the late 1960s. Irrigators were required to maintain groundwater monitoring wells and report regularly, as per schedule, a practice that still continues and enables sustainable management to be maintained (in 2005, reports were received for 168 of the 181 registered monitoring wells (ABWMC, 2005)). In this way data is generated twice a year, enabling managers to monitor trends to guide water allocation and planning.

The most important initiative, however, was to replace a substantial proportion of the groundwater allocations with abstraction from the Murray River system. Allocations taken from groundwater were cut by 30 per cent, and irrigators who converted to lake water could avoid these cuts. In this way there was a 'voluntary' shift of irrigators to lake water from the Murray River system, and the conditions were developed to allow the replenishing of groundwater and local water tables. As a result, groundwater extraction now represents only a small fraction of total allocations (DWLBC, 2007).

Water resources management plans and regulation

Water Allocation Plans (WAPs) now regulate water use from four prescribed water resources in the Board's catchment, including the River Murray Prescribed Watercourse and the Angas Bremer Prescribed Wells Area. Policies relating to permits for water extraction regulate activities that might impact both prescribed and non-prescribed resources.

The Angus Bremer WAP goes back to 1981, 20 years before the Murray River WAP, and served as a model for other WAPs in the state. The WAPs have become more sophisticated over time and address more dimensions of water management than simply quantity. Thus the Murray–Darling WAP addressed not only quantitative allocations but also considered dependent ecosystems, water resources, capacity of resources to meet demands, water allocation criteria, water transfer criteria (including water rights, licensing and permits) and monitoring, as well as transfers to related irrigation management zones (including Angas Bremer).

The WAP process also instils discipline amongst water users. The Angus Bremer regulation regime calls for strict compliance and reporting on groundwater and use of irrigation water from all sources, and accredits irrigators annually. For example, 110 irrigators were accredited for the 2005–2006 irrigation year, while 23 were excluded because they did not complete the Irrigation

Annual Report forms or because of late lodging of forms.

This attention to discipline has been rewarded in that the South Australian Murray-Basin Natural Resources Management Board recognizes that the Angas Bremer region has demonstrated effective water-use efficiency and should continue to be managed separately from the remainder of the Murray River system (SAMBNRMB, 2007).

Water-use efficiency planning and practice

South Australia's Angas Bremer irrigation district has attracted national and international interest for its leadership in planning for water-use efficiency, and then implementing it through strict regulations and good technical practice.

Efficient water use is vital to avoid the ever-present threat of waterlogging and salinization. An innovative common technology, the CSIRO FullStop wetting front detector, has been adopted by all irrigators to tell them when to stop irrigating. Each irrigator has installed monitoring wells used to measure and report the height and the salinity of the water table.

But efficiency is also driven by simple economics. So the shift from local groundwater sources to water 'imported' from the Murray River, which required investment in pipelines and pumping and significantly increased the cost of water to the farmers, was an important contributor to improved efficiency.

In 1992, four local farmers commissioned a privately-owned water scheme to provide supplies to a vineyard, market garden, lawn turf business and lucerne farm. Water was pumped through 10 kilometres of 375 millimetre PVC pipe from Lake Alexandrina, delivering up to 160 litres per second.

Three years later, the State Premier opened the largest, privately funded, irrigation scheme in Australia. The A$2,600,000 (approximately US$2.1 million) scheme takes water from Lake Alexandrina through 35 kilometres of pipeline. It has 40 participants, can pump 470 litres of water per second, and supports a variety of production activities including grapes, almonds, potatoes, lucerne, dairy and horticulture, as well as stock and domestic needs.

Many more schemes were subsequently implemented, putting into practice the shift from groundwater to surface resources. Since the users have to carry both the capital and operating cost of these schemes, they have every incentive to use water productively and efficiently, and this has been reflected in the changing patterns of production.

Agreed code of practice and environmental stewardship

Angas Bremer is a prime example of the development of partnerships as an instrument for promoting IWRM goals.

In 2001 the public–private partnership measures in Angas Bremer produced a much acclaimed Angas Bremer Code of Practice for Irrigators, which has instructions in simple language and addresses conjunctive water-use issues. The Code of Practice has four major components:

- quarterly monitoring of groundwater levels;
- monitoring of irrigation application volumes and irrigation drainage at each irrigation event;
- planting and nurturing of a prescribed area with deep-rooted perennial vegetation; and
- reporting through an annual reporting scheme.

Under this Code, irrigators commit themselves to applying the FullStop technology to avoid over irrigation, as well as to maintaining ground cover to retain appropriate rates of evapotranspiration throughout the year. The Irrigation Annual Reporting scheme enables irrigators to report on their compliance with the Code and their licence conditions under the Angus Bremer Water Allocation Plan (ABWAP). The state government conducts 10 per cent random audits to check that the self-audits are robust.

The Code is complemented by the Land and Water Management Plan, and provides the basis for the Watermark Environmental Stewardship Programme of the Murray–Darling Basin Commission. The Angus Bremer community is keen to develop the Code into an accredited EcoLabel scheme with which to market their products. A water module is also in place as part of a new Environmental Management System for grape farmers.

The success of Angas Bremer's participatory and largely self-governing compliance system is attributed to the following factors:

- long-term access by the community to technical support;
- partnerships among irrigators, industries and government agencies based on mutual trust, and understanding of the regional and local natural resources and their associated risks;
- strong leadership within the community, local industries and government agencies;
- support for community groups by paid project officers;
- a core group of committed local volunteers representing a broad range of irrigation interests; and
- a viable, profitable industry that allows irrigators to have the resources to do more than merely survive.

In addition, the geographical features of the ABPWA are such that any impacts of inefficient irrigation will be visible on-farm or on neighbouring farms within one generation (Muller, 2002: 2).

The experiences in Angas Bremer have influenced larger-scale participatory programmes linking good water practices with trade and industry. The Environmental Management in Viticulture initiative has run trials in the entire Langhorne Creek area to establish the Angas Bremer Code of Practice for managing irrigation at Langhorne Creek (see www.angasbremerwater.org.au/files/ Trial%20report.pdf). Funded by the Murray– Darling Basin Commission under the Watermark programme, the objective of the trial is to develop a regional framework for improving environmental stewardship offering tiered levels of recognition linked to catchment targets. This is a variation of the traditional ISO1400 style Environmental Management System (EMS).

The programme includes a methodology for assessing the environmental risks of viticultural activities, the determination of a series of environmental performance targets that both the region and individual growers can aspire to achieve, locally developed environmental standards, and severe performance and processed based steps or 'levels' designed to give recognition for participant's efforts. Best Management Practices are developed with the support of local Natural Resource Management (NRM) officers to address environmental issues identified by the growers. The requirement of annual review provides for continuous improvement.

The cap on water extractions and environmental flows

Perhaps the most significant management instrument introduced more recently has been the 1995 decision of the Murray–Darling Basin Commission to limit or cap the annual levels of water extraction from the Basin rivers to the levels that existed in 1993 and 1994.

This not only confirms that the Basin is effectively 'closed' and that there is no additional water for new uses, but also establishes the foundations for actually reducing existing allocations to provide for environmental requirements. Under this decision, while annual water extractions across the Basin will continue to vary with climatic conditions, the average of these extractions remains constant thereafter.

The cap does not affect the shares of water from the transboundary Murray River for the three states in the Murray–Darling Basin. There remains a monthly minimum volumetric entitlement for the downstream state (in this case South Australia), from which the Angas Bremer farmers draw their surface

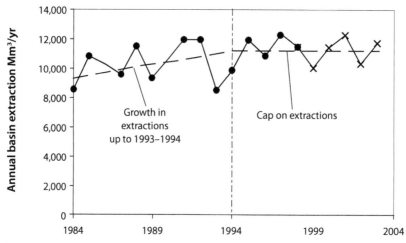

Source: Haisman, 2004: 61

Figure 7.5   Operation of the cap on water extractions

water allocation. Allocation to individual water users and between consumption and in-stream uses remains a matter for the State concerned. What the cap does is to reinforce the responsibility of the states to exercise controls over actual extractions of water in accord with the cap rules. The cap, by sending a powerful signal that water is finite, is intended to improve water-use efficiencies.

On the downside, the cap works only on annual flows, ignores the major extraction or intervention caused by the operation of storage dams, and allows maximum extractions in dry years (which may be contrary to the actual ecological needs of the rivers at that time). Nonetheless, the cap has demonstrated through strict implementation that managing the health of rivers is important (Haisman, 2004: 64).

Water rights and water trading

The cap has also reinforced the status of water entitlements as an important instrument of management. While the concept of water entitlements or rights has been inherent in Australian water policies, it takes on particular significance in the new dispensation. This is highlighted by the objectives of the National Water

Initiative, which, as explained by the National Water Commission, signifies:

- a commitment to identifying over-allocated water systems, and restoring those systems to sustainable levels;
- the expansion of the trade in water resulting in more profitable use of water and more cost-effective and flexible recovery of water to achieve environmental outcomes;
- more confidence for those investing in the water industry due to more secure water access entitlements, better registry arrangements, monitoring, reporting and accounting of water use, and improved public access to information; and
- more sophisticated, transparent and comprehensive water planning.

Tradable water rights are a means of 'moving' water to where it may be most productive in economic terms. But in a stressed situation such as the Murray, they have to address a great deal more than simply a quantitative allocation. The Murray–Darling WAP and its review in 2007 speak of water rights covering the following aspects:

- water access entitlement (a right to access the water resource);
- allocation (a share of the consumptive pool);
- water resource works approval (right to install and use water extraction infrastructure);
- site use approval (right to use water at a particular site); and
- delivery capacity entitlement (used to manage access in a constrained system).

In the Angas Bremer area, the existence of these entitlements made it possible to shift production patterns from the traditional focus on lucerne hay to wine grapes. Licensing helped access necessary water, and some farmers were able to sell their water rights to others (including some ranchers selling rights to wine grape growers). In addition, the increased asset value of a water entitlement has enabled an increased financial investment in water-efficiency technologies by individual water users. It also encourages efficiency since the nature of Australian water-use rights is such that a water efficiency gain made by the holder of the right does not diminish the quantum of the right, but enables the holder to use the water saved to increase production (Haisman, 2004: 63) (in many other countries, water 'rights' are measured in terms of the area that can be irrigated).

However, a new dimension was added to Australia's water market when interstate water trade was enabled. Tagged trade of water between states in the Murray–Darling Basin was set to commence from July 2007 as per the requirements of the National Water Initiative. The WAP review of 2007 has called for a process where there is the ability to attach rules to the traded water (so that, for example, any rules managing salinity impacts attached to an existing water right would be preserved in the trade).

Communications and information management system

Communications and information management is vital for effective water management in complex circumstances such as those in Angas Bremer. In addition to the Angas Bremer Code of Practice, the Irrigation Annual Reporting system provides a simple, low-cost process through which each irrigator reports information about metered water use, water table levels and cropped areas, and receives collated and analysed data for the whole region.

The Angas Bremer Water Management Committee (ABWMC) communicates regularly with its irrigator members. It produces formal annual reports and public meetings are held each year to consider the findings and take necessary action. It also maintains a website on which much of the information is available (www.angasbremerwater.org.au/maps.htm), including:

- The Angas Bremer Land and Water Management Plan;
- Angas Bremer Soil Book;
- Angus Bremer Map Layers CD (with over 100 map layers); and
- Geographical Information System.

This constant communication with irrigators has contributed to the effective management of the local water resources.

Institutions

Effective water management depends on the establishment of effective institutions, as is well demonstrated in the Angas Bremer case. Although there had been many ad hoc efforts to address particular problems, the establishment of the ABWRC was an important milestone.

It was the ABWRC that tackled the fundamental problem of historic overallocation of groundwater and initiated the shift to using surface water from the Murray system. Once integrated into the Murray system, the Angas Bremer irrigators became part of the River Murray Catchment Water Management Board (RMCWMB), which superseded the ABWRC in 1997. However, the community retained its ability to act collectively by electing a voluntary committee, the Angas Bremer Water Management Committee (ABWMC), which entered into a formal partnership with the RMCWMB to develop and implement the Water Allocation Plan (WAP) for the Angus Bremer region.

The committee also works closely with the South Australia Department of Water, Land and Biodiversity Conservation. This tripartite partnership approach has ensured that innovative policies can be developed and implemented in a technically robust manner while remaining under community ownership.

While this 'alphabet soup' of organizations is confusing, it accurately reflects the complex mix of particularities and interdependency that often characterize water management efforts in large and complex river basins. In this respect, the needs of the Angas Bremer irrigators could not have been met were it not for the evolution at national and basin level of institutions that created a cooperative framework within which they could find their niche.

Outcomes

What began in the Angas Bremer Irrigation District as collective local action by a small group of irrigation farmers working with government officials to solve their highly specific local water management problems has, over a few decades, become inextricably linked to water reform on a national level. By participating in these reforms and, arguably, providing some leadership to them by their practical example, the farming community has avoided what could have been a crisis of viability and sustainability, and this community has expanded substantially.

Physical and economic outcomes

In Angas Bremer both the farm gate income and the area of irrigation increased, while the volume of irrigation water used decreased. This was achieved

by implementing new technologies, by changing the crop mix to take advantage of emerging market opportunities and by cooperation between users to implement the changes.

A measure of these changes is given in Table 7.1, which shows how water sources and volumetric water use evolved over a 20-year period that marked a doubling of the irrigated area.

Table 7.1 *A comparison of ABPWA groundwater and Murray River water use (volumetric and area) in 1981 and 2001*

	1981	2001
Groundwater use (GL)	26.6	1.5
Recharge (GL)	6	2–6?
Murray River water use (GL)	Negligible	16
Total irrigation water use (GL)	26.6	17.5
Irrigated area (ha)	3,400	6,800
Lucerne Hay (ha)	c.1700	471
Total area water use (ML/ha)	7.8	2.6
Vines area water use (ML/ha)	n/a*	2.1

*n/a = not available; Irrigation Annual Reporting did not commence until 1996
Source: ABWRC (1996) and ABWMC (2001)

The data show a reduction in total water application across the region from 7.8 megalitres per hectare in 1981 to 2.6 megalitres per hectare in 2001. Over that 20-year period, a 95 per cent reduction in groundwater use was achieved, stabilizing the groundwater situation and allowing the pressure in the confined aquifer to return to its pre-irrigation state, while the water table showed a steady rise.

The reduction in water use reflects a change both in the crops grown and in increased water-use efficiency. Such changes have been associated at the national level with substantial increases in the commercial productivity of water. In vegetable production, the return per megalitre of water is reported to have increased from A$1,762 (US$1,400) in 1996/1997 to A$3,207 (US$2,500) in 2000/2001 (Sylvia and Skewes, 2008). This has been attributed to the increased use of water-

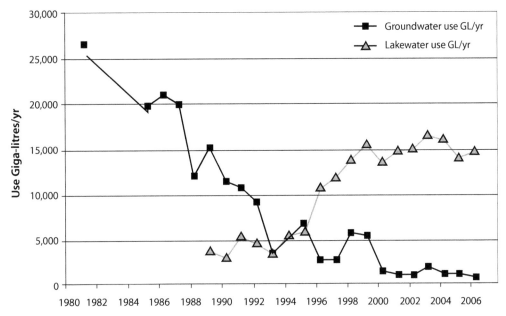

Source: DWLBC, 2007

Figure 7.6   Annual groundwater allocation and use in the Angas Bremer PWA

efficient delivery systems such as drip irrigation, increased use of recycling on-farm, wide-scale adoption of irrigation scheduling and soil moisture monitoring, and increased use of whole-farm planning and soil mapping (Sylvia and Skewes, 2008). Many of these initiatives had been adopted in the Angas Bremer District.

However, the most important contributor to the Angas Bremer success story has been viticulture, driven by a booming wine trade. This transformed the economics of irrigation in the region and allowed many of the investments needed to address the water resource challenges to be made. Were it not for the opportunities created by the growth of the wine industry and the branding of the surrounding Langhorne Creek as one of the original *terroirs* of Australia, the outcome might have been very different.

The establishment of an Angus Bremer Code of Practice and its linkage to the Environmental Stewardship Programme is an interesting and important illustration of the way in which better management practices can be supported and enhanced by partnerships with consumers and other stakeholders. The focus on raising standards of vineyard-based environmental management contributed to higher incomes and environmental sustainability, as well as providing further incentives for good water management.

Social outcomes

The outcomes of the Angas Bremer case are less clear in terms of its social dimensions. The social impact of water trading is controversial since it has not been clear to whom the benefits will flow. Certain politicians were deeply suspicious that moving large volumes of water through paper trades would ultimately hurt both farmers and consumers. Others, including Sharon Starik of the South Australian Farmers Federation, believed water trading offered a good business opportunity and source of revenue for farmers wishing to sell their unused water rights. And still others, such as

Amy Goodman, a manager at the River Murray Catchment Water Management Board, indicated that 'on average, there are not huge amounts of water to transfer anyway. It is not causing significant movement of water' (Commonwealth of Australia, 2004).

However, outside of South Australia, it was clear that trading was impacting the broader community as opposed to individual farm owners. A recent report on the issue in neighbouring Victoria State concluded that:

> Communities can find change and adjustment difficult. The communities in the case study regions that were exporting water experienced reduced populations and less local spending. Communities in the case study regions importing water experienced increased populations but did not necessarily have the infrastructure and services to accommodate the new arrivals.
>
> (Frontier Economics, 2007)

In Angas Bremer it has been suggested that part of the reasons for long-term success is the nature of the community and its social capital: the Angas Bremer Irrigation District contains only around 160 growers, and many have lived in the area for generations. It has also been suggested that the strong local football club has helped further the sense of community among the irrigators and their families (Thomson, 2004). Thomson also comments that:

> Angas Bremer growers do not have the future option to profit and retire after selling their land to grow houses because the district is too far away from Adelaide. Growers are keenly aware that their livelihoods and the livelihoods of future generations depend on the sustainable management of their resources so that agriculture continues to be viable in the long term.

While the trading issue remains contentious, it was overtaken by the impact of the 2007/2008 drought that left many farmers without the water required to sustain their crops. The fact that there was a mix of annual and perennial crops, of which the annuals could be sacrificed to preserve the perennials, has helped some of the irrigators to weather the challenges to date. However, this response has been cited as a contributor to a global rice shortage in 2008 (Bradsher, 2008).

Lessons learned

The Angas Bremer case demonstrates that achieving goals of improved water management requires long-term vision, time and patience. It also illustrates the importance of the economic framework within which water is managed. Thus the high prices of wine grapes encouraged the switch from lucerne hay to grapes. The nature of the community that allowed people to commit themselves to achieving their goal, combining their resources and 'dreaming the same dream' as they work together, was also clearly a contributor to the area's success.

A further lesson is that in a federal state such as Australia, it may be necessary to achieve effective national coordination over the management of interstate rivers before local challenges, such as those that face the Angas Bremer community, can effectively be dealt with.

In the end, the successful adaptation of the Angas Bremer farmers was only possible because of the overall water management framework provided by the Murray–Darling Basin Agreement and the institutions that it gave rise to. Local action could work, to turn water into wine, because that broader framework was in place.

References

ABWMC (Angus Bremer Water Management Committee) (2001) Angas Bremer Code of Practice, Australian Water Environments, Natural Heritage Trust, Issue No. 1

ABWMC (Angas Bremer Water Management Committee) (2005) *Irrigation in the Angus Bremer Irrigation Management Zone: 2004–2005 Annual Report, 2005,*

Angas Bremer Regional History, Angas Bremer Water Management Committee, website: www.angasbremerwater.org.au

Bradsher, K. (2008) 'Drought in Australia, a Global Shortage of Rice', *The New York Times*, 17 April 2008

Commonwealth of Australia (2004) *Senate, Rural and Regional Affairs and Transport References Committee, Reference: Rural water usage in Australia*, Tuesday 20 April 2004

DWLBC (2007) *Angas Bremer PWA Groundwater Status Report December 2007*, Department of Water, Land and Biodiversity Conservation, Government of South Australia, Adelaide

Frontier Economics (2007) *The Economic and Social Impacts of Water Trading – Case studies in the Victorian Murray Valley*, Report for the Rural Industries Research and Development Corporation, National Water Commission and Murray–Darling Basin Commission, RIRDC Publication No. 07/121

Haisman, B. (2004) *Murray–Darling River Basin Case Study Australia*, Background Paper, (produced for) World Bank research project: *Integrated River Basin Management and the Principle of Managing Water Resources at the Lowest Appropriate Level – When and Why Does It (Not) Work in Practice?* World Bank: Agriculture and Rural Development Department, Water Resources Management Group and the South Asia Social and Environment Unit, Washington

Muller, K.L. (2002) *A Partnership Approach to Environmental Stewardship in Langhorn Creek, SA*, paper presented at the Second National Wine Industry Environment Conference and Exhibition, Adelaide, SA, November

SAMBNRMB (South Australian Murray-Basin Natural Resources Management Board) (2007) *River Murray Water Allocation Plan Review*, Government of South Australia, Adelaide, June

Schonfeldt, C. (2000) *Future Water Resources for South Australia*, ATSE Focus No. 111, March/April, www.atse.org.au/index.php?sectionid=440

Sylvia, S. and Skewes, M. (2008) *Maximising Returns from Water in the Australian Vegetable Industry: South Australia*, New South Wales, Department of Primary Industries, pp1–22. Also available from www.ruralsolutions.sa.gov.au

Thomson, T. (2004) *Learning together at Angas Bremer, SA Department of Water Land and Biodiversity Conservation*, paper presented to Irrigation Association of Australia Conference in Adelaide, 12 May 2004

Part Two – Basin Level

A number of the local cases, notably those of Sukhomajri and the Angas Bremer area, show that, while very local water management challenges in small communities can successfully be addressed by local action, there are limits to such action. All too often, action in one community impacts on others, both positively or negatively, and a broader framework of action is required.

This is a critical message. While the principle of subsidiarity, dealing with issues that only affect local people at local level, is important, it should not be interpreted to mean that local issues can be solved at local level solely within a local framework.

The challenge is to know when and how to make the connections with the larger universe. So while, initially, the Angas Bremer farmers could manage their limited groundwater resource entirely locally, to sustain and expand their farming activities they had to engage in the broader challenges of managing water in the much larger mother basin.

The implication is that arrangements at the larger scale must provide a robust framework for big, strategic, decisions, while allowing local issues to be managed locally. In many cases, the appropriate level for this is the river basin.

The establishment of an effective basin-wide institutional framework that still allows and encourages local-level cooperation is arguably one of the greatest challenges of water resource management and is often an evolutionary process. In Japan's Lake Biwa, the tension between local, regional and national priorities was at the centre of the conflicts that continued for many years before some harmony was achieved. In the Yangtze case, on the other hand, it can be argued that, while the 'macro-management' framework is in place, the linkages from national to basin to local level are still emerging.

In this context, the Lerma–Chapala case is encouraging, representing as it does another case in which a process of trial and error has led to an outcome which, if not perfect, has addressed the most acute challenges and provided a platform for a sustainable future for the local residents.

8

Turning Water Stress into Water Management Success: Experiences in the Lerma–Chapala River Basin

Jorge Hidalgo[1] and Humberto Peña

River basins around the world are experiencing high levels of water stress, with water withdrawals greatly exceeding utilizable levels. The Lerma–Chapala in Mexico is an extreme example. In this, one of the world's most over-committed basins, demographic growth, coupled with industrial and agricultural development, led to a serious imbalance between water withdrawals and availability. Upstream agricultural drain of the Lerma River contributed to continual declines in water levels and the shrinking surface area of Lake Chapala, while water quality decreased steadily as the demands from human settlements, farmers and industry continued to grow.

The dying lake and the fierce competition for water set the stage for concerted action by various stakeholders. A regional basin council, Mexico's first, with input from three levels of government and representatives from various water user groups, was identified as the best way to move towards sustainable water management. Individual states agreed to assume responsibility for clean-up programmes and a master water plan was created with input from stakeholders, including civil society. There are now local Basin Councils, local Water User Authorities and local Aquifer Management Councils throughout the region.

The case shows that heightened hydrological, social and ecological interdependencies in river basins, caused by increased pressure on water resources, demand more integrated management approaches. And it also suggests that adaptive, multilevel, collaborative governance arrangements may be better suited to cope with competing political and economic interests and increasing water scarcity than more centralized forms of water management.

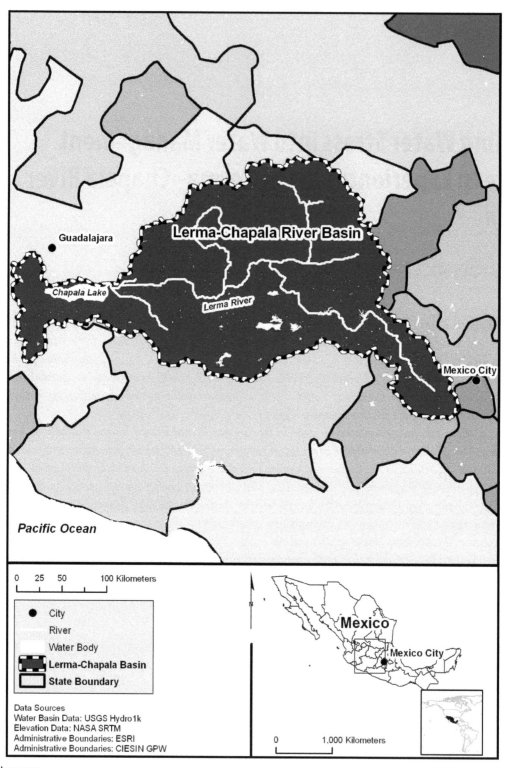

Location map

Mexico's move from development to integrated management

At the beginning of the 1970s, Mexico presented its first National Water Plan. The plan was the basis for changing the approach to water management in Mexico – from water resources development to integrated water resources management (IWRM). The conclusion of this important policy document was that Mexico had enough water and land to support its future development, but only if those resources were used and managed in an efficient way (CPNH, 1975). The Plan put forward nine recommendations for water resources management to ensure its contribution to wealth and welfare:

- To make use of irrigation and drainage infrastructure at maximum efficiency.
- To assure that all the regions with hydroelectric potential were developed.
- To build pilot projects as the first steps to large actions where experience or information were insufficient.
- To give an impulse to and motivate formal and informal capacity-building of human resources at all levels.
- To stimulate and organize appropriate basic and applied research on Mexican water realities to improve the use, reuse and control of water quantity and quality.
- To promote permanent education and public

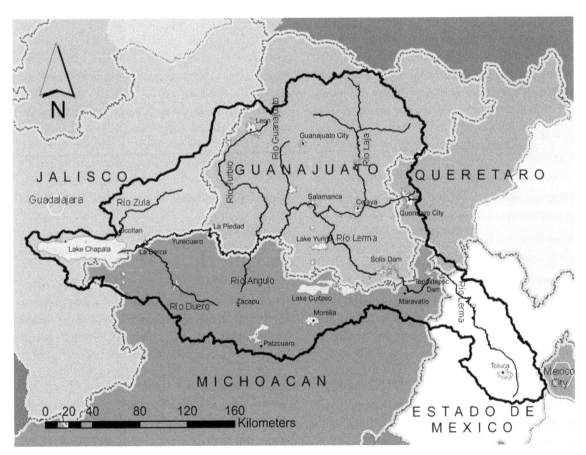

Source: Wester, 2008

Figure 8.1   The Lerma–Chapala River Basin

awareness programmes to create a new water culture for the rational water use.

- To promote stakeholders' participation in the planning, construction and operation of hydraulic works.
- To implement integrated programmes of infrastructural and institutional actions and support.
- To take into account in planning and management decisions the interdependence of surface water and groundwater as a same resource, at the local and regional level.

The Plan was approved by Mexico's President in 1975, and the next year the National Water Plan Commission was established to implement, control and evaluate it. To realize the recommendations it was necessary to change the approach to water governance. So at the beginning of the 1980s, six regional water resources offices, dependants of the Water Resources and Agriculture Secretary, were installed to study and get better information on the river basins that they covered. This was the beginning of a new organization in the water sector of Mexico.

One of these regional water agencies was the Lerma–Chapala River Basin Regional Management. It was given responsibility for planning the water resources of the Lerma–Chapala Basin. It is important to know that Mexico is a republic integrated by the union of 32 states, so there are three levels of government: the federal, the state and the municipal. Each state and municipality is independent of the federal government, but, despite this, historically the country has had a very strong centralism. Thus the decision of the federal government to create these new regional agencies was an important first step towards decentralizing the water administration and achieving more efficient coordination with the states and municipalities.

The Lerma–Chapala Basin is part of the hydrological system formed by the rivers Lerma and Santiago (see Figure 8.1). It includes part of the states of Guanajuato, Jalisco, Mexico, Michoacán and Queretaro, and has a surface area of 51,887 km². The Lerma River originates in the Chignahuapan lagoon, in the State of Mexico in the higher part of the western Sierra Madre; it has a length of 708 km and ends in Lake Chapala. At this point the Santiago River originates and, after traversing 650 km, ends in the Pacific Ocean.

Since the early 1980s water has not flowed naturally from Lake Chapala into the Santiago River, due to the intensive use of water in the middle and lower reaches of the Lerma. Thus the Lerma–Chapala system is considered a closed basin from a hydrological point of view, with no natural outflow. The mean annual precipitation in the region is 771 mm, but most of it is concentrated in the south, in the highland areas, while in the centre and northern part of the basin the precipitation is lower, between 400 and 500 mm. This precipitation regime gives to the region a semi-arid climate with rains in the summer.

The Lerma–Chapala Basin is located in what was historically Mesoamerica, and for that reason its natural resources have a long history of exploitation. However, it was not until the 20th century, and specifically after the 1940s, that gradually an imbalance in the use of water began to grow – driven by the massive construction of hydraulic infrastructure, high demographic growth, the industrialization process and a large increase in irrigated agriculture.

In 1940, there were 2.5 million inhabitants, of which more than half lived in rural areas. Between 1950 and 1990 there was a period of rapid demographic growth, with an annual mean population growth rate of 2.65 per cent. By 2000 the river basin was supporting around 11 million people living in the basin (10 per cent of the country's population), plus another 5 million people that benefited from the transfer of its water to supply the cities of Guadalajara and Mexico City. For the year 2030 the projection of the population is estimated at 13.2 million inhabitants inside the basin.

The relation between the population and the available water indicates that there is a strong water stress in the basin; this indicator is approximately 1,000 m^3/inhabitant/year. As a result, competition between water uses is becoming more intense and the conflicts surrounding water allocation, use, quality and conservation are increasing in frequency.

There are six large cities with more than 1 million inhabitants that are competing for water in and outside of the basin: Toluca-Metepec-Lerma, Leon, Morelia and Queretaro inside, and Guadalajara and Mexico City outside of the hydrologic region. Also there are 34 localities with more than 20,000 inhabitants; this number of cities concentrated in the basin is greater than in any other basin in the country. Among the urban population, 94 per cent has access to a domestic water supply, 91 per cent to the sewer systems and just 32 per cent to the sanitation systems. In the rural areas, 76 per cent of the population has access to water supply, 38 per cent to sewerage and only 1 per cent to sanitation. According to incomes, half of the population has a low socioeconomic level, 30 per cent a medium level and 20 per cent a high level.

Between 1980 and 1990, rapid industrial development occurred in the basin. At present there are several large industrial corridors that combined generate almost 66 per cent of the gross national product of the country's industrial sector, equivalent to around US$74 million. Although these industries do not demand much water, they compete for a scarce resource and discharge a high concentration of pollutants to the river and water bodies (see Figure 8.2).

Agriculture also experienced a period of rapid growth. More than half of the land in the basin is used for agriculture – approximately 3 million ha, of which 830,000 ha are equipped with irrigation infrastructure. The most important irrigation projects were built from 1940 to 1980. This sector uses 92 per cent of the surface water and 75 per cent of the groundwater available in the basin. The global efficiency of the irrigation districts is low; it varies between 33 and 40 per cent, but at basin level water-use efficiencies are very high due to the reuse of water.

The environmental degradation is a clear sign of poor water and land management in the basin. Due to its geographical location, the volumes of water stored in Lake Chapala are an indicator of the behaviour of the total watershed. In particular, the lake reflects the growth of water demand in the upper part of the basin and also the effects of wastewater discharges without previous treatment that have degraded the quality of the lake's water. In the 1990s and early 2000s, the Lerma–Chapala Basin passed through a serious crisis whose more notorious manifestation was the sharp decrease in the volumes of water stored in Lake Chapala (see Figure 8.3), the largest natural lake in Mexico and the third largest in Latin America. Whereas the Aral Sea lost 75 per cent of its volume in four decades, Lake Chapala lost 90 per cent of its volume in two decades, from 1981 to 2001. In July 2001 the lake held only around 15 per cent of its capacity, and, adding insult to injury, this water was severely contaminated. Some water managers supposed that this drop in the storage of the lake was due to drought conditions in the region, and certainly drought was a factor, but it was not enough to explain such a dramatic water shortage. Other mentioned causes were the growth of the demand due to demographic and economic reasons, but if the government's data are reviewed, the direct water extraction from the lake in the 1990s was not that different from, for instance, the 1980s. Rather, this crisis was mainly caused by historical deficiencies in water governance in the basin. Due to accelerated economic growth and lack of good data, between the 1940s and 1980s the federal water authority gave an over-concession of the water rights. Not surprisingly this led to over-exploitation of water sources and a severe water imbalance (see Figure 8.4). Between 1991 and 2001, Lake Chapala received less inflow than the amount of water leaving the lake through evaporation and extractions for Guadalajara and irrigation. This

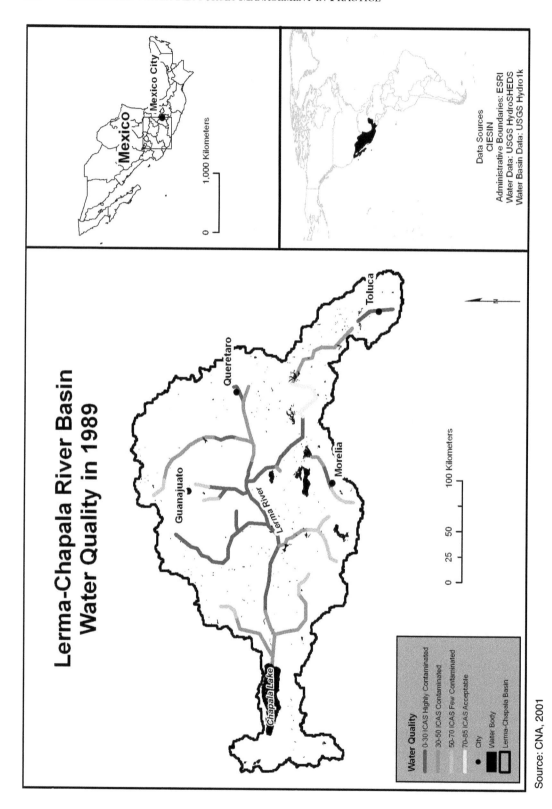

Source: CNA, 2001

Figure 8.2   Water quality in the Lerma–Chapala Basin in 1989

yielded an annual mean deficit of 400 million hectare metres (hm^3).

The mean precipitation in the watershed generates a runoff of 5,513 hm^3 and an annual aquifer recharge of 4,016 hm^3 in the 37 aquifers identified. Irrigated agriculture is the main water user in the region: eight irrigation districts and 1,136 small irrigation units deplete some 4,550 hm^3 of surface water (this includes 830 hm^3 of evaporation from all the dam reservoirs and water bodies in the hydrologic region, except for Lake Chapala, which evaporates 1,440 hm^3 per year), while almost 12,000 small irrigation units use 3,869 hm^3 groundwater. The second main water user is the urban sector; with Guadalajara using 237 hm^3 and Leon, Morelia and Toluca together some 60 hm^3 of surface water (Dau and Aparicio, 2006). The rest of the localities satisfy their demand through almost 3,000 deep wells that exploit 841 hm^3 of groundwater; 35 per cent of this volume is transferred to support the demand of Mexico City. The total water depletion of these three uses (irrigated agriculture, urban water and evaporation from water bodies) is already more than the renewable amount of water available. If we include the other water users, the overexploitation of surface water resources is 982 hm^3 and the groundwater deficit is 1,143 hm^3. Therefore, of the 37 aquifers identified in the region, 18 are over-exploited. Static water levels vary between 40 m of depth in the equilibrium aquifers and more than 120 m of depth in the over-exploited aquifers.

In addition to water over-exploitation, water pollution and soil degradation are serious issues in the basin. The environmental impacts of agricultural and industrial activities are severe water pollution and land degradation covering 72 per cent of the basin's area. The two most important processes are declining soil fertility (accounting for 57 per cent of land degradation) and soil erosion (accounting for 33 per cent of land degradation). The highest rate of negative land-use change is found in the forest areas; deforestation in the upper basin is a serious problem that has to be addressed.

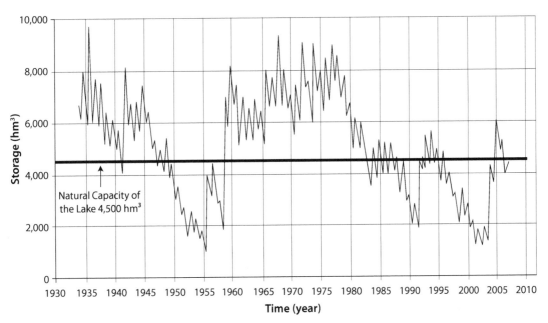

Source: Guitrón et al, 2003

Figure 8.3   Lake Chapala water storage evolution

As a result of these factors (fast economic growth, high demographic density, and poor water management), the water-use and environmental problems in the Lerma–Chapala basin have increased since the late 1970s, making it difficult to sustain economic development and at the same time to preserve the environment, especially where it concerns Lake Chapala. At the time of writing, the basin has the following problems (CNA, 2001):

- considerable reduction in the water level of Lake Chapala between 1981 and 2003, despite the existence of a Surface Water Distribution Agreement for the basin signed in 1991, which could not reverse this trend;
- groundwater overexploitation of nearly 1,143 hm³/year;
- water quality deterioration due to point and diffuse water discharges estimated at 49 T/day of nitrogen and 15 T/day of phosphorus;
- low irrigation efficiency, estimated at 35 per cent, in almost 800,000 hectares distributed within irrigation units and districts;
- low average efficiency in urban water supply systems, estimated at 44 per cent, and low water supply coverage in rural areas;

- growing conflict between water users due to competing water uses;
- degradation of water and other natural resources due to lack of environmental consciousness;
- high social, cultural and productive heterogeneity, which makes natural resource management very complex;
- lack of coordination between user groups and governmental offices responsible for natural resources management; and
- a poorly functioning hydro-climatologic measurement and monitoring network.

The approach

As discussed above, starting from the 1980s the natural conditions in the basin changed radically; the hydrological cycle and the ecosystems of the basin were seriously altered by human activities, and at the same time a drought period started. Lake Chapala began to dry up (see Figure 8.3) and water scarcity began to increase. The conditions for unsustainable development and water conflicts were established.

To address this increasingly dire situation the federal government established the Lerma–Chapala Basin Regional Management (LCBRM). This was the federal government's first step towards admitting that centralized water management was not an appropriate approach and that changes were needed in the institutional roles.

During the 1980s the LCBRM worked hard in order to collect more information, improve the water plan, define better institutional roles and, most importantly, to involve all basin stakeholders in decision-making. The results from this effort were manifested in the first coordination agreement between the federal government and the governments of the five states that share the water resources of the basin, which was signed in 1989. The purpose of the pact was to modify water allocation mechanisms, to improve water quality, to

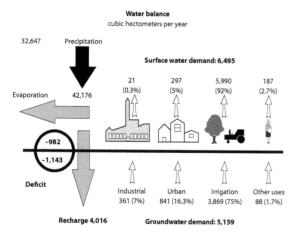

Source: CNA, 2001

Figure 8.4   Water balance of the Lerma–Chapala Basin

increase water-use efficiency and to conserve the basin's ecosystems.

To implement and evaluate the actions derived from the agreement it was necessary to establish a Control and Evaluation Advisory Council (CEAC), a technical working group consisting of representatives of the federal and state governments. The CEAC, which was formed in September of 1989, designed new rules for surface water reallocation. These were approved by consensus by the main stakeholders and signed in August of 1991. At the same time, the first stage of the water treatment programme had begun with important works in the most crucial areas of the basin. There were 106 water treatment plants planned under the programme, which were capable of treating 14,821 l/s of effluents discharged to the Lerma River.

The experiences gained in the Lerma–Chapala Basin influenced changes in the 1972 Mexican Federal Water Law. In December of 1992 the Federal Water Law was reformed and important modifications were included. One of them was to include provisions for the constitution of advisory and coordination bodies that can formulate and execute actions and programmes to improve water management, develop hydraulic infrastructure and preserve water resources – i.e. the Basin Councils.

The new law established the rules of the game for an enabling environment. Thus, based on the 1992 National Water Law (NWL), the CEAC was transformed into the Lerma–Chapala Basin Council (LCBC) on 28 January 1993. With the creation of the first basin council in Mexico, the water policy in Mexico began to build new mechanisms and forums for enabling all stakeholders to play their respective roles in the development and management of water resources, and for facilitating and exercising their participation. On that date, the second stage of the sanitation programme started along with several other important actions, such as the promotion of fishing, clean-up, efficient water use and watershed management programmes.

As part of the decentralization promoted by the federal government, in December 1989 the National Water Commission was established as the federal water authority. During the 1990s, the irrigation districts were transferred to the agricultural users. In each state a State Water Commission was set up as a facilitator for the water authority, and the water users were organized in civil associations in order to participate as representatives of their water uses in the Basin Councils.

More recently, on 30 April 2004, a revised version of the National Water Law was published, based on the following principles:

- integrated water resources management is the better way to approach water conflicts and its problems;
- water resource and its management is considered as a strategic resource and national security; and
- the basin as the unit of planning and management.

The revised water law further clarified the enabling environment, defining institutional roles and management instruments as complementary elements of an effective water resources management system; it strengthened the Basin Councils, giving them more responsibilities and an organizational structure; and it defined the basin commissions, the technical watershed committees and the technical aquifer committees as instances of participation at different levels. It also transformed the river basin regional management offices into basin organizations, recognizing them as the regional water authorities. At the time of writing, the Lerma–Chapala Basin Organization (LCBO) and the LCBC share responsibilities for leadership in water governance in the basin.

Instruments used

In order to tackle the problems of the river basin, and within the context of the institutional changes described above, diverse instruments were developed. These can be grouped under the following points.

River basin water resources management plan

The first instrument developed was the Lerma–Chapala Basin plan. With this instrument the authorities and stakeholders could prioritize their problems and strive for the best solutions for everybody based on win–win scenarios. The priorities were:

1   to define a new mechanism to reallocate the water for all uses and users in the basin;
2   to clean-up the water in the whole region;
3   to improve the efficiencies in the hydraulics systems, and
4   to restore as much as possible the damaged environment.

The plan was developed with the participation of the stakeholders, discussed with government representatives and agreed to by consensus. Thus, its implementation was easy; everybody cooperated and allocated sufficient financial means to implement the two stages of the water treatment programme.

Programme of water resources assessment and its use

Water resources assessment was another important instrument used. The assessment, which determined water resources availability and demand, was critical for defining new rules for water reallocation. Over the course of two years, the appropriate hydrological studies were conducted and the results were presented and approved by the CEAC in August of 1991. The water accounting showed that the natural water system in the basin had been changed drastically; there was no more available water in the river basin, and both surface water and groundwater were revealed to have a significant deficit (see Figure 8.4).

The agriculture users demand more water than any other users, and, before the 1991 Agreement Act, nobody took into account the environmental water demand for Lake Chapala. Thus, the rules to reallocate the water considering Lake Chapala as a user were relevant and an advanced tool at the time. Unfortunately, the disposable data were not enough to define efficient rules for allocating the water in a drought period; and so during the period 1994–2001 the level of Lake Chapala went down 4 m, 70 per cent of the volume was lost, and the conflict among water users increased.

This situation spurred the LCBC to hire the Mexican Institute for Water Technology (IMTA) to perform scientific studies and develop appropriate technology tools. The first step undertaken by IMTA, in order to improve the water balance, was a hydrological study of the lake, with an emphasis on the proper estimation of evaporation and the runoff in watersheds without gauge systems. After that, it made an assessment of water pollution in the basin and the lake, and some studies to determinate the water consumption, the public perception of water value, and to analyse the social conflicts over water.

Technical support for the agreements

A special tool was developed by IMTA to understand why the allocation rules did succeed in restoring Lake Chapala and to soften the conflict over water. IMTA developed a dynamic basin simulation model, together with a special technical group that represented the interests of the stakeholders in the different states. All together worked hard:

1   to understand the complex systems (natural and human), they studied how the key elements of the systems were related and interacted;
2   to build the model, they chose a simple software platform (Stella® software);
3   to calibrate the model, it was necessary to be transparent with all data and information used;
4   to validate the model, several tests were

conducted in order to gain trust in and outside of the group; and

5   to design and evaluate scenarios, each member of the group simulated the scenarios that best represented the interests of its state and stakeholders, and could appreciate the impact of the decision.

The group evaluated all the scenarios using economic, social, technical, political and environmental criteria. The model proved to be essential in the consensus building process, since it helped to create the proper climate for discussions based on facts rather than opinions; thus it was possible to overcome the existing impasse.

This exercise was very helpful to sensitize all participants in the process of conflict resolution and negotiation that followed. IMTA acted as a mediator in the process until the new agreement was signed on 14 December 2004 by the President of Mexico and the five State Governors. On 14 January 2005 it was signed by the users' representatives in Guadalajara, Jalisco.

The negotiation process lasted more than three years and more than 35,000 working hours were invested in it by the participating parties. The new 2004 Water Allocation Agreement considers hydrological, economic, social and environment criteria and uses a dynamic basin simulation model, together with an optimization model based on a genetic algorithm, to decide how to allocate the water to all users, including Lake Chapala.

Communication and transparency policy

Special attention has been given to communication and information systems. These instruments have been used by the LCBO and LCBC from the 1980s onwards. The LCBC is the main forum for communicating what happens in the basin. All decisions are made by the Basin Council, and it is responsible for summoning and informing all the stakeholders about the advances achieved in relation to the basin management plan. The community is well informed by all kind of media and the LCBO has an online information system at www.cna.gob.mx/LermaWeb.

The Red Lerma was recently established to help both basin organisations in decision-making and information dissemination. This network is led by the six main universities in the basin. They share research projects oriented towards providing specific solutions to the relevant problems identified in the basin and they contribute to institutional capacity-building.

Development of the economic and regulatory instruments

One of the causes of the conflicts in the basin has been the over-concession of the water rights. The solution to this problem was conceived in terms of combining two management instruments – allocation by a market-based instrument in conjunction with a regulatory instrument. As there is no extra water available in the basin, it is imperative to design public policies and instruments that promote efficient water use, order the concessions, regulate the existing water market operations and facilitate the transfers of water rights, warranting the fulfilment of the law to the benefit of the public interest and the basin. The reform of the 2004 National Water Law contemplated the establishment of water banks that would regulate the transfers of water rights, based on clearly defined rules. The proposal made by IMTA was to implement the Lerma–Chapala Water Bank. This bank will be a central element of public policy to solve the water scarcity problem in the basin in the short, medium and long term, because it will help to regulate the supply and demand of the water resource through the voluntary transfer of water rights between users.

The Lerma–Chapala Basin and IWRM

The efforts carried out in order to solve the problems of the Lerma–Chapala Basin have been inspired by

IWRM. In effect, the solutions were based on an integrated vision that took into account the diverse elements that directly or indirectly are associated with water resources management.

Integrating water management within the administrative division of the country and ensuring water was on the political agenda were important first steps. But to achieve these, a change in awareness was needed – people had to realize that water was at the centre of a larger crisis (economic, social and political) that threatened the set of activities in the river basin and that gave water a strategic importance for development.

The success in achieving this change was reflected in the signature of the Surface Water Distribution Agreement in 1991, in the results of the first stage of the cleaning programme, in the actions of reforestation of the most damaged zones of the river basin, and in the creation of the groundwater technical committees and of the Turbio River Basin Commission, which governs the most contaminated tributary of the Lerma River.

Implementing IWRM is not an easy task; it requires a great effort of persuasion, patience and perseverance, which means time. Because the implementation of IWRM depends on, first, the good will of politicians and establishing the favourable conditions of an enabling environment, second, the organization of institutional roles, and, third, identifying and implementing the appropriates management instruments, it takes a long time to see outcomes and impacts. In Mexico it has taken more than 30 years to see some results, but at the beginning of the 21st century we are satisfied that we have made the right decisions at the right times.

The Lerma–Chapala Basin has been the IWRM laboratory in Mexico. Important outcomes of the lengthy negotiations that have taken place in the basin are the signing of three agreements acts: the first was signed in 1989 to achieve the water resources and sanitation plan in order to preserve Lake Chapala and reach a sustainable development in the basin; the second was signed in 1991 to coordinate the allocation and uses of the surface water; and, finally, the third was signed in 2004 to coordinate the recovery and sustainability of the Lerma–Chapala Basin.

The outcomes and impacts

The implementation of the tools and actions described above has had several impacts.

From the social point of view, the stakeholders know more about the problems of the basin; they are sensitized to the issues and are more willing to participate in the solutions, and, as a consequence, conflicts are minimized. The LCBC has been consolidated; its new organizational structure allows more flexibility to attend to the problems and conflicts related to the basin, sub-basins and aquifers. It plays the main role in creating popular awareness and understanding, and it has been the facilitator to ensure collaboration across sectors and boundaries. The Red Lerma is another good indication of the will to participate: the universities took the initiative by themselves to coordinate and share experiences and knowledge to solve important problems, to develop appropriate technologies, to build institutional capacities, to disseminate information and to advise the LCBC.

Environmentally speaking, Lake Chapala recovered its natural capacity level before the last agreements act. A volume of 955 hm^3 was released from the dams to the lake, which is now considered another user in the water allocation decision-making. The new allocation rules require that the water surplus, in excess of the dams' operation levels, have to be released to the lake. The water quality of the Lerma River is improving. At the end of the second stage of the water treatment programme, it had cleaned up 8,952 l/s (60.4 per cent) of all water discharges in the river. With the signing of the 2004 Agreement Act, the LCBC approved the beginning of the third stage of the water treatment programme, which would cover 100 per cent of all discharges.

Technological advances have also benefited the basin. The dynamic river basin simulation model in conjunction with an optimization model is one example. There are other models to study the aquifers in the basin: for example, the Guanajuato State Water Commission has developed groundwater models to analyse its overexploited aquifers. The irrigation districts of the states of Mexico, Michoacan and Jalisco have signed agreements with the National Water Commission to modernize their irrigation systems in order to improve their efficiency. The water saved will be deducted from the water rights of these irrigation districts, and will be reallocated for the sustainability of the basin and the recovery of Lake Chapala.

The economic impacts that came from the agreements can be estimated in the irrigated agricultural production. The new water allocation rules guarantee that irrigated agriculture receives at least 50 per cent of the maximum volume conceded by federal government during drought seasons; in the past agriculture received at most 30 per cent of conceded volumes during droughts. Also with the cooperation of all stakeholders, it was possible to finance the two first stages of the sanitation programme: US$240 million was invested in construction of 52 water treatment plants.

The change in the National Water Law related to basin organizations is another important legal impact of the experiences with IWRM in the Lerma–Chapala Basin. The influence of the LCBO and LCBC is generating benefits not just in Lerma–Chapala, but also in the rest of the country. It is in the Lerma–Chapala Basin that several important experiments and changes in water governance occurred, in such a way that they will transform institutional roles throughout the country. Those initiatives are now part of the approach of integrated water resources river basin management in Mexico. The current water policy promotes the decentralization of water resources management with a river basin focus, and with a broad participation of the stakeholders and governmental authorities in order to reach sustainable water resources management.[2]

Maybe the most important challenge in Mexico is how to finance all the activities required by IWRM. It is clear that the core of IWRM is the compromise and the cooperation of all the actors in the solution of the problems, but in Mexico the main financial source is federal funds. The integration of local governments and private or social-capital contributions with the federal donation is not enough to cover all the solutions. So, we have to be intelligent and efficient at times in applying the budget to the river basin development programmes.

Key lessons learned

From the experiences with IWRM in the Lerma–Chapala Basin we can list some key lessons (Mestre, 2005).

Institutional lessons

- In Mexico it was possible to implement integrated water resources management at the desirable level of the river basin. We adopted a bimodal water management model – looking for a natural equilibrium between the authorities and stakeholders, and, in the case of the Lerma–Chapala Basin Organization and its Basin Council, the model has been working very well.
- It was very productive from the beginning to consider the three government levels in the negotiations with stakeholders.
- The agreements signed considered not only the political commitments, but also the human, material and financial resources to realize the water resources basin programme.
- It is very important that any agreement has to have a corresponding financial backing, as in that way it is easier to gain the support of the population.

- The General Assembly of the LCBC is a fundamental part of the basin governance; it is a democratic and transparent platform for stakeholders' participation. Both governments and civil society must collaborate in the prevention and resolution of conflicts. It is clear that the goals are more easily achieved when all groups are working together.

Social lessons

- Society responds faster than government; centralism is waning, and the demands of society are changing the way the authorities are perceived and the way government works.
- The private sector is increasingly taking the lead in development projects.
- Local authorities are empowered by IWRM processes.
- Well-organized civil organizations respond strongly during decision-making processes, and are able to share responsibilities.
- The river basin commission is a good forum for discussions of diverse visions and for decision-making. It encourages tolerance between interested parties and opens the way for processes of participative democracy.
- It is urgent that society be enabled to increase its knowledge and ability to engage productively in debates; the construction of a new water culture is indispensable in the advancement of the IWRM.

Notes

1. Since 1995, Jorge Hidalgo has served as researcher with the Mexican Institute for Water Technology, which provided technical support to the Lerma–Chapala process described here.
2. The dynamic participation of this Regional Management (the LCBRM) in bringing a new vision to bear on the problems of the basin was one of the reasons the International Network of Basin Organizations had its first general assembly in Mexico; Eduardo Mestre was Manager of the LCBRM and, at the same time, he was president of INBO in 1994.

References

CNA (Comisión Nacional del Agua) (1988) *Diagnóstico de la Región VIII: Lerma-Santiago-Pacífico*, Subdirección General de Programación, Mexico

CPNH (Comisión del Plan Nacional Hidráulico) (1975) *Plan Nacional Hidráulico*, Comisión del Plan Nacional Hidráulico, Mexico

Dau, F.E. and Aparicio, M.J. (eds) (2006) *Acciones para la recuperación ambiental de la cuenca Lerma–Chapala*, Comisión Estatal de Agua y Saneamiento, Jalisco

Guitron, A., Hidalgo, J. et al (2003) 'A water crisis management: The Lerma–Chapala Basin case', in Brebbia, C.A. (ed.) *Water Resources Management II*, United Kingdom, Wessex Institute of Technology, WIT Press, pp345–354

Mestre, E. (2005) notes from the course *Integrated Water Resources Management in River Basin*, Instituto Mexicano de Tecnología del Agua, Jiutepec

Wester, P. (2008) *Shedding the Waters: Institutional Change and Water Control in the Lerma–Chapala Basin, Mexico*, PhD dissertation, Wageningen University

9

Turning Conflict into Opportunities: The Case of Lake Biwa, Japan

Simi Kamal

Freshwater lakes are often a focus for interaction among upstream and downstream communities and those who live around the lake. Lake Biwa in Japan is a typical case, with a relatively small population upstream (engaged in intensive agriculture), the populous megacities of Kyoto, Osaka and Kobe downstream, and a substantial community living around the lake whose economy has gradually shifted from a focus on manufacturing industry to tourism.

Reflecting this interaction, the history of Lake Biwa's management is one of conflicts over competing uses of water – between water for nature and water for development, between the local administrative unit of the Shiga Prefecture and the central government, and between rural watershed areas and the downstream megacities.

This case is an example of the evolution of water resources management, from an initial focus on infrastructure development – aimed at supporting economic development and protecting communities against floods – to a focus on protecting water quality, and finally to an approach that seeks to meet the needs of a diverse group of stakeholders while achieving environmental sustainability and more efficient water use.

The development context and water challenges

Formed about 4 million years ago, Lake Biwa is an ancient lake of great historical and cultural significance. With a surface area of 670 km², Lake Biwa is the largest lake in Japan. It lies in the upper reaches of the relatively small (8,240 km²) Yodo River Basin, above one of the most urbanized and developed regions in the world (Nakamura, 2007: 34).

While more than 400 tributaries flow into the lake, only one natural watercourse, the Seta River, flows out of Lake Biwa. The Seta River is joined by the Kizu and the Katsura Rivers to become the Yodo River, which flows into Osaka Bay and eventually to the Pacific Ocean (see Figure 9.1).

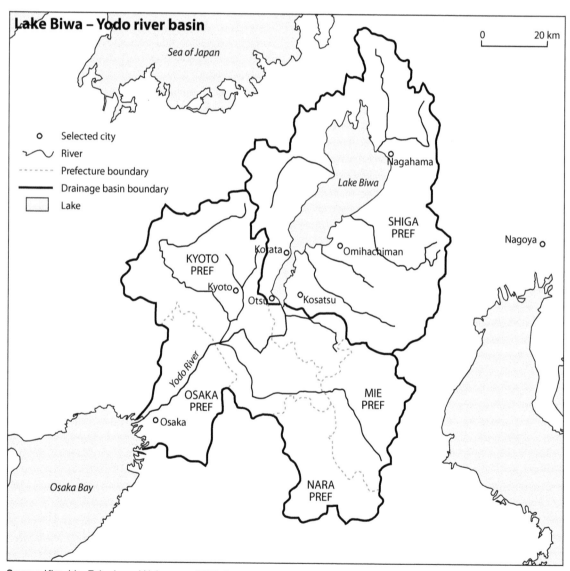

Source: Kira, Ide, Fukada and Nakamura, 2005: 1

Figure 9.1   Lake Biwa and Yodo River Basin

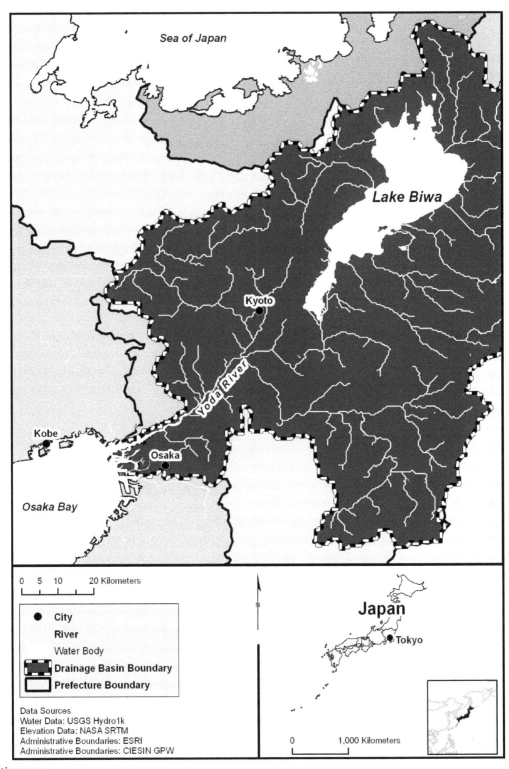

Location map

The history of Lake Biwa's planned management goes back more than a century to an episode of severe flooding in 1896, which caused substantial damage in the region. This led to the dredging of the Seta River at the outlet of the lake and, soon after, the construction of the Seta weir, the lake's first artificial water-flow control facility.

Half a century later, Lake Biwa became an important focus of development to support the programme of industrialization and urbanization that was initiated after the Second World War. It was the main source of water to meet increased demands from the burgeoning industries and urban areas of the Kinki region (including the cities of Osaka and Kobe).

By the 1960s, it became apparent that the explosive industrial and population growth in the region was leading to wide-scale pollution and the destruction of important habitats around the lake. In addition, extensive urbanization and development saw wetlands and connected lakes and waterways filled in to create new agricultural lands. One consequence was deteriorating water quality, as a result of increased nutrient loading and depleted oxygen levels. In addition, important environmental habitats were being lost (Lake Biwa Museum, 2008):

- 85 per cent of the area of attached lakes and wetlands, which are crucial for the spawning of some of the endemic and commercially important fish, were filled in as rice fields.
- Over 50 per cent of reed beds – an important habitat for birds, fish and invertebrates, which also help to remove excess nutrients from the lake – were lost.
- Extensive channelization of inflowing streams of rivers impoverished the fauna and flora.
- Land reclamation and development saw approximately one-third of the lake's shore built up, with major roads running along the lake.

During the 1970s, pollution became so bad that it began to threaten both human and ecosystem health. Chemicals from agricultural runoff, untreated sewage and wastewater, and industrial effluents, including heavy metals, combined to degrade soil and water quality and contaminate fish and shellfish. Excessive levels of nutrients such as nitrogen and phosphorous caused massive algal blooms and red tides in the lake in 1977. The eutrophication – the reduction of dissolved oxygen in the water due to overgrowth of algae and other plants – encouraged invasive species to flourish in one of the most biologically diverse areas of Japan (Kira et al, 2005).

The problems were compounded by further human intervention when alien species, especially black bass (*Micropterus salmoides*) and blue gill (*Lepomis macrochiru*) were introduced into the lake for sport fishing and dramatically increased in number, at the expense of the native fish.

In addition, the regional Shiga Prefecture government actively promoted the construction of infrastructure along a substantial portion of the lake – much of which was designed to support tourism.

Responses – addressing the challenges

While the history of water resources development interventions in Lake Biwa goes back to the 1950s and before, the conflicts began to be critically articulated in the 1970s. From this time onwards there were a number of processes with IWRM characteristics that were tried and that led to a cascade of results over three decades (see Table 9.1).

Infrastructure interventions

Initial infrastructure development was to address the flooding and water shortages that threatened communities downstream and around the lake. According to official records, the 1896 flood that provoked the earliest interventions inundated 16,594 hectares of land around the lake for eight months, as water levels rose 3.7 metres above

Table 9.1 *Lake Biwa water resources development and conservation milestones*

Year	Laws, regulations, institutions
1950s	Water resources development interventions in Lake Biwa
1961	The Water Resources Development Promotion Law
1963	Kinki Region Improvement Law (KRIL)
1967	The Basic Law for Environmental Pollution Control
1969	Pollution Control Ordinance, Shiga Prefecture
1970s	Heavy contamination in Lake Biwa
1970	Environmental Standards for Water Quality
1970	Start of the citizens' movement in Shiga Prefecture
1971	Water Pollution Control Law
1971	Pollution Control Ordinance, Shiga Prefecture
1972	Law for Lake Biwa Comprehensive Development (LLBCD)
1972	Lake Biwa Comprehensive Development Project (LBCDP)
1976	Start of litigation by citizens of Osaka and Kobe
1977	Red tides in Lake Biwa
1977	Citizens' Movement focuses on water quality issues
1978	Citizens' Forum for Conservation of the Aquatic Environment around Lake Biwa (Biwa-ko Forum) formed
1980s	Continuation of litigation and public outcry
1981	Shiga Environment Conservation Association formed
1989	Court rules against litigants
1990s	Dams built around Lake Biwa
1990s	Environmental Cooperatives formed
1992	Reed Belt Conservation Ordinance, Shiga Prefecture
1993	Lake Biwa designated a Ramsar site
1996	Basic Environment Ordinance
1997	The LBCDP increases water flow to 40 metric tonnes per second
2000s	The concept of the 'mother lake' takes root
2005	Lake Biwa Renaissance Plan

normal. This led to the dredging of the outlet and the construction of the Seta weir to enable outflows and lake levels to be controlled.

But too much water was only one aspect of the problem. In drought years, downstream communities dependent on flows from the lake also faced water shortages when there was not enough outflow. In 1978, water restrictions were in place for 161 days of the year – with serious economic and social impacts for downstream communities. In addition, navigability of the river, at that stage still important for fishing boats, was impeded by shallow water during dry periods (Japan Water Agency, 2003).

To address these issues, a further series of infrastructure interventions was initiated in 1972 through what became known as the Lake Biwa Comprehensive Development Project (LBCDP). This actually consisted of two main projects, the Lake Biwa Development Project led by the Water Resources Development Public Corporation (what is now the Japan Water Agency), and a Regional Development Project overseen by the government administration at the national, prefectural and local levels.

The most important infrastructure work under the LBCDP allowed navigation and water abstraction to be maintained even when the lake levels fell. Extensive flood protection works were built and new drainage pumps were installed to lift flood water over the new flood levees and prevent inundation in the area around the lake.

Lobbying for water quality and environmental protection

The red tide incident proved to be a wake-up call for the residents around Lake Biwa at a time when there was growing consciousness about the potentially devastating impact of water pollution – in other nearby areas, heavy metal contamination had poisoned people through consumption of fish from polluted water. The type of algal growths that affected Lake Biwa was potentially poisonous and, at the very least, made it difficult to produce drinking water with an acceptable taste and smell. Detergents, fertilizers used by rice farmers upstream, untreated sewage and industrial discharges were all contributing to the problem of eutrophication in the lake.

People were alarmed and there were two noteworthy citizens' actions in the 1970s – a people's movement in the upstream areas and one of Japan's first environmental lawsuits to be brought by citizens, in this case by downstream water users.

The citizens' movement had its base in the Soap Movement, which started in the early 1970s in the area around Lake Biwa as a campaign by homemakers who were concerned about babies' nappy rash and housewives' eczema caused by synthetic detergents. In 1977, however, after the red tides in the lake, the movement changed its focus to the conservation of lake water quality and became one of the most successful and celebrated citizens' movements in Japan (Kira et al, 2005). This movement is especially significant because it was led by women, who highlighted that citizens were also responsible for the degradation of lake water quality.

The movement successfully put pressure on the Shiga Prefecture government to pass the Lake Biwa Ordinance, to regulate the use of phosphorus-containing detergents. This presaged a worldwide trend to reduce the use of phosphorus-based detergents in catchment areas vulnerable to eutrophication.

The second citizens' action was led by residents of downstream Osaka, who filed a lawsuit in 1976 against the central government and the Shiga Prefecture government for supplying them with polluted water. While the case was lost, it established the fundamental argument for the quality of water and water conservation.

Instruments used

Achieving a balance between development and conservation in Lake Biwa required an 'integration' of economic concerns with environmental sustainability. It also entailed adapting to the changing social priorities of groups in the upstream and downstream areas. This process involved several instruments, which are considered classic tools of IWRM.

Development and implementation of the water resources knowledge base

The development of the water resources knowledge base and the capacity to use it was a key early intervention needed to guide other interventions. Lake Biwa is reputed to be one of the most researched areas in Japan, with a rich scientific and social database related to water resources. For example, there are long records of physical data such as flood levels and rainfall, and environmental information such as limnological data is available from 1965 to date (Ueda et al, 1998). Thanks to strong ties to academic institutions, this knowledge base is continuously being maintained and updated.

In addition, capacity has been developed across many disciplines with a whole cadre of trained researchers, scientists and field staff now working in the area. The Lake Biwa Environmental Research Institute was established in 1993 as a mechanism for gathering, exchanging and distributing research information and technologies which reflect the needs of society and challenges of the government. The Centre for Ecological Research (Kyoto University), the UNEP-supported International Environmental Technology Centre and the Lake Biwa Museum are also centres of research. The outcome is a permanent system of close linkages between research institutions and citizens' groups, private businesses, industries and government institutions, such that the water resources knowledge base continuously feeds into management decisions.

Public participation and citizens' action

The early 1970s was a critical time for the evolution of environmental consciousness in Japan. The period saw a steady succession of legal actions against polluters, resulting in victories in the four major pollution trials (Minamata Disease in Niigata, Yokkaichi Asthma, 'Itai-Itai' Disease and Minamata Disease in Kumamoto). The findings in favour of the victims prompted a revision of environmental standards and compensation plans, and caused a fundamental shift in thinking on pollution – it went from being considered an acceptable price for economic growth to being considered generally unacceptable (IIC-JICA, 2005).

Campaigns against pollution spilled over into the broader community, and public participation and citizens' action were an important instrument of social change around Lake Biwa, helping to build consensus as well as to encourage water-use efficiency. The Soap Movement was one of the most visible examples of this and, with the participation of a wide range of organizations, spread throughout the Shiga Prefecture. But the Soap Movement was just one element of broader public action. A Citizens' Forum for Conservation of the Aquatic Environment around Lake Biwa (Biwa-ko Forum) was established in 1978 and remains relevant today as an established institutional outcome of citizens' action. Another outcome of this movement was the establishment of Environmental Cooperatives in 1990, which specialized in the promotion of

environmentally sound commercial products, especially those that end up in the drainage system after use.

Private initiatives

The participation of all stakeholders, including, critically, the private sector, is a key element of IWRM. In the context of Lake Biwa, there was a unique instance of private initiative when in 1981 the Shiga Environment Conservation Association was formed. It consisted of more than 400 local companies at that time and originated from an information exchange group of personnel in charge of industrial wastewater treatment. This association has remained an active participant in the conservation and reuse of water. Other private organizations serving Lake Biwa include the Citizen Forum for the Conservation of the Aquatic Environment around Lake Biwa (Lake Biwa Citizen Forum), Friends of Lake Biwa, Water and Culture Study Group, Akanoi-Biwako Environment Citizens' Initiative, Lake Biwa-Yodo River Water Purification Organization and Environmental Co-op Union Shiga. By adopting codes of practices these organizations have directly contributed to the outcome of improved water quality.

Wastewater treatment and sewerage

While many of the large infrastructure projects were challenged by the citizens' movements, there was recognition that local sources of pollution had to be addressed. When the LBCDP received a ten-year extension in 1982, a significant amount of investment was budgeted for environmental components. Investment in sewerage and night soil treatment was increased by ¥26 billion (US$200 million) from 1982 to 1991. New project components such as dairy waste management, agricultural community sewerage, refuse disposal, and a surveillance and monitoring system, were also added. In addition, new technology was introduced to manage waste from isolated households and small communities.

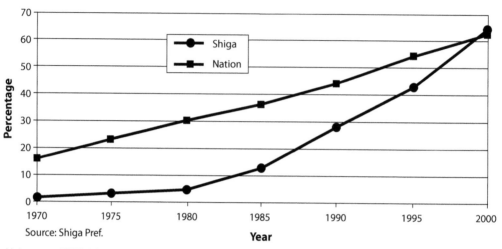

Source: Shiga Pref.

Source: Nakamura, 2007: 14

Figure 9.2   Percentage of population served by sewerage systems in urban areas, 1970–2000, Shiga Prefecture, Japan

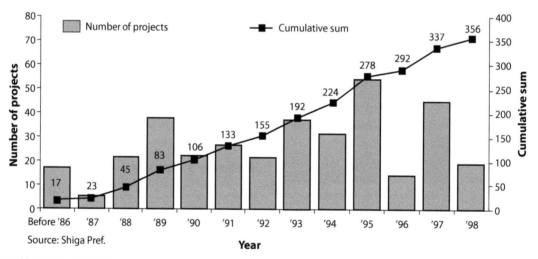

Source: Shiga Pref.

Source: Nakamura, 2007: 14

Figure 9.3   Coverage of rural community sewerage, 1986–1998, Shiga Prefecture, Japan

Conflict resolution through managing disputes and sharing water

In its early stages, along with their benefits, the LBCDP and related regional developments also triggered massive destruction of the lakeshore and littoral ecosystems, resulting in the degradation of lake water quality.

This triggered another set of responses, this time from downstream residents. In 1976, more than 1,000 citizens, mostly residents of Osaka, initiated a lawsuit against the central government and the

Table 9.2 *Changes in the environmental administration system of the Shiga Prefecture government*

Year	Departments and divisions in charge of Lake Biwa and its catchments area
1970	Department of Welfare (Antipollution Measures Office)
1972	Department of Planning, Life Environment Bureau (Antipollution Division, Drinking Water and Waste Management Division, Nature Conservation Division, Prefecture Life Division)
1974	Department of Life Environment (Antipollution Division, Environmental Policy and Waste Management Division, Nature Conservation Division, Prefecture Life Division)
1979	Department of Life Environment (Environment Office, Waste Management Division, Nature Conservation Division, Prefecture Life Division)
1996	Department of Lake Biwa and the Environment (Water Policy Administration Division, Environmental Policy Division, Waste Management Division, Ecological Lifestyle Promotion Division, Nature Conservation Division, Forest Conservation Division, Forestry Administration Division, Sewerage Planning Division)

Source: Kira et al, 2005

Shiga Prefecture government to stop the project, as their drinking water supply was becoming polluted. While, at first glance, the litigation appeared to be about the demands of the downstream users to curtail development of water resources for upstream uses, it was actually about controlling water pollution and ensuring high-quality water for all uses and users.

The plaintiffs lost the case after 13 years of civil lawsuit, but the fundamental argument for the quality of water and water conservation came to fruition. Many of the concerns expressed about ecosystem integrity in the suit not only turned out to be justified, but also correctly implied the direction of environmental policy in the post-project period.

As the concerns of the upstream watershed inhabitants also shifted from comprehensive development to comprehensive conservation of Lake Biwa, they also become an instrument of change and integration. Although the citizens' movement in Shiga Prefecture (upstream) and the litigants of Osaka (downstream) had different interests, their goals converged, which provided a window for converting conflict into opportunities.

Building effective institutions and administrative systems

The challenge of addressing water quality and environmental conservation required substantial organizational changes in the Shiga Prefecture government to reflect the changing focus and priorities towards environmental conservation. The evolution of the organizations involved in environmental protection in Lake Biwa between 1970 and 1996 is shown in Table 9.2. This evolution demonstrates the slow shift from a focus on antipollution measures in 1970 to a more comprehensive menu in 1996, where 'ecological lifestyle promotion' is also a function.

These changes have now produced a structure that reflects the multiple concerns of the Shiga Prefecture in respect to Lake Biwa.

Legislation for water quality and environmental conservation

The national legislative framework that currently governs Lake Biwa took many years to develop. The framework takes an IWRM approach – attempting to 'balance' development with conservation. It covers the management of water quality and nature under environmental conservation, river channel improvement and the development of catchment forests as means of flood control, quality water resources for both upstream and downstream areas and the improvement of fisheries. The process of addressing the challenges faced around Lake Biwa contributed substantially to the evolution of this national-level framework.

The national Basic Law for Environmental Pollution Control was enacted in 1967. The Shiga Prefecture government went a step further than the national level legislation: it set stricter standards for industrial wastewaters and developed its own Water Pollution Control Ordinance (1969) for its jurisdiction. But even this was not sufficient, as was demonstrated by the freshwater red tide in 1977. As a result, the Eutrophication Control Ordinance was enacted by the Shiga Prefecture in 1979, prohibiting the use, sale and gift of synthetic detergents containing phosphate, and setting the very first nitrogen and phosphorus standards for industrial effluent in the world. Six years later the national Water Pollution Control Law was revised, setting down effluent standards of nitrogen and phosphorus for all Japanese lakes.

In 1984 a national law for Lake Water Quality Conservation was enacted under which each local government in charge of a designated lake has to formulate a water quality conservation plan every five years (Kira, 2005: 66). While Lake Biwa is just one of ten lakes designated under the national 'Clean Lakes Law', it was the legislative work of the Shiga Prefecture that led the effort in improving water quality of lakes in Japan.

In 1992 the Shiga Prefecture Government enacted the Reed Belt Conservation Ordinance (the first act in Japan advocating the importance of an ecosystem) and the Basic Environment Ordinance in 1996. These actions gave effect to good intentions that had been expressed earlier: Lake Biwa was designated a quasi-national park in 1950 and the entire Lake Biwa region was designated as a wildlife sanctuary in 1971. Further progress was made when Lake Biwa was registered with the Ramsar Convention on Wetlands in 1993 as a wetland of international importance.

Financing and incentive structures

The LBCDP, which eventually expanded to cover a 25-year period (1972–1997) with a budget of ¥1.8 trillion (US$18 billion), has been an extremely costly venture, and the inclusion of the environmental protection and conservation elements substantially increased the costs. A key element in the rehabilitation and transformation of the lake has thus been the development of financial mechanisms to support the process. Critically, these included the mobilization of substantial contributions from both national government and downstream beneficiaries of better lake management under the guidance of local stakeholders.

There are few charges for water in Japan, except tariffs for tap water and wastewater removal. Most public works, including the construction of major sewerage works, lack financial mechanisms to recover the expenses and therefore depend on allocations of public funds. This reflects a situation in which public funds are allocated through political processes, and public works are promoted for job creation, local economic development and sometimes to address the demands of specific interest groups.

Part of the challenge in Lake Biwa was thus to channel extensive public works funding into interventions that were guided by local priorities rather than national sectional interests.

Outcomes

Lake Biwa's is a case where citizens' actions helped to promote better water resource management. As a direct result of citizens' actions, the LBCDP was drastically revised and lake biodiversity protected (and eventually restored). While the LBCDP was originally focused on resolving the water supply problems of downstream users, the scope of the project eventually expanded to include flood control, water-level control, irrigation and agricultural development, forestry, fisheries and nature conservation.

Industrial interests responded to pressure and innovated to address public concerns, notably through the introduction of new generation detergents nationwide, a trend that quickly had global impact.

Reducing the negative impacts of floods and droughts

By 1997 the LBCDP had achieved its goal of being able to provide reliably 40 m³/s of water to downstream users even in a drought period (Kada, 1999), although this capacity is not currently fully used. In the dry year of 1994, after the project became operational, water restrictions were imposed for only 44 days, although lake levels were the lowest for more than 20 years. Ten years before, restrictions due to a less acute drought had lasted for 156 days. The expansion of levees along the lake greatly reduced flooding risk, as did the Seta River's increased discharge capacity. As a consequence, the area of land flooded in recent wet periods was only 740 ha for 11 days, compared to over 4,000 ha for twice as long in 1961 and 16,600 ha for 238 days in 1896 (IAAJWA Lake Biwa Development Integrated Operation & Maintenance Office, 2003).

Before the interventions, agriculture in the Biwa Basin suffered from both water shortages and flooding; 90 per cent of the 600 km² of irrigated rice paddies now use reliable water supplies from the lake and are largely protected from flooding, although there are still no economic incentives to save water (Kira, 2005: 65) and users have to meet only their direct costs of supply.

The outcome of all of this is that Lake Biwa is efficient and integrative in its physical functioning. It is a control basin during time of flood, protecting communities, and a reservoir in times of drought, providing water security and highlighting the key objective of good water resource management – supporting economic activity, social equity and environmental sustainability.

Water quality

While the new infrastructure has been relatively successful in addressing the physical challenges, the broader environmental rehabilitation process has proved to be more difficult. The measures adopted succeeded in reducing nutrient loading from domestic and industrial sources, which contributed to the revitalization of the lake and its ecosystem. One unexpected consequence of the regional development efforts, however, was that the intensification of rice production and the switch to irrigation supplied by pumping from the lake (rather than by a cascading system of paddies fed from local sources) led to an increase in diffuse pollution from fertilizer runoff.

Urban development also saw an upsurge in 'diffuse pollution', which is more difficult to control than point-source pollution from easily identifiable activities. Thus conservation efforts and development activities continued to be at odds throughout the 1990s.

While the measures put in place seemed to have improved the water quality to some degree – or at least managed to halt the upward trend in pollution – water quality indicators such as chemical oxygen demand, total phosphorous and total nitrogen are still above the national environmental standards in most parts of the lake (Shiga Prefecture, 2008).

Economic impacts

The cleaning up of the lake, water conservation and improved water quality has also had many direct and indirect economic impacts. Not least is that the lake has become a great tourist attraction with over 37 million people visiting each year – this tourism supports the economy of the area (LakeNet, 2003).

The annual volume of freshwater fish went up to about 2,800 tons and, in 2004, accounted for 50 per cent of nationwide sales of home-grown freshwater fish, although more recently fluctuations and weather impacts attributed to climate change have negatively affected the industry.

Future challenges

By the beginning of the 21st century, the concept of the 'mother lake', extolled by the citizens' movement, had been adopted by planners and government, making water conservation the joint and prime concern of both government and citizens for the Lake Biwa area and the Shiga Prefecture. This culture of respect for the natural environment bodes well for the future of the lake.

However, many challenges remain. The Lake Biwa Research Institute has warned that global warming has already led to the deterioration of the living environment in the lake – affecting aquatic life adversely. It has concluded that lower snowfall in the area has resulted in an oxygen-depleted zone in the deep water layers (Kira, 2005: 62). Dead fish, lower spawning rates and other impacts have also been documented (Masaki, 2008). Conservation concepts now have to include adapting to the impacts of climate change, highlighting the importance of adaptive strategies of which IWRM is a generic example.

In addition, Lake Biwa's water quality problems are far from over. The management of Lake Biwa water quality is about to enter a new phase, which will accelerate point- and nonpoint-source control measures and introduce measures to achieve a greater degree of lake ecosystem integrity. Lake quality is to be further improved in terms of parameters other than chemical oxygen demand, total phosphorous and total nitrogen. The emerging features of control measures characterizing the new phase include the realignment of protected watersheds and land uses, development of ecotone areas – transition zones between ecosystems – including restoration of the once reclaimed attached lakes, and the integrated management of priority watersheds.

The realignment of the protected watershed and rezoning will probably involve an extensive political process. According to Nakamura (1998):

> The major issue will be to devise proper economic incentives such as compensatory payments for existing land owners, as almost every piece of watershed land is owned and engaged in some sort of productive activity … As for ecotone development, agricultural and urban sectors will have to collaborate closely on infrastructure development or redevelopment around coastal regions of the lake and numerous watercourses. The key to integrated management of priority watersheds is to redesign the watercourses hitherto independently developed by different sectors of the government.

The Lake Biwa Renaissance Plan has been developed with the concept of 'Water Connects People, Nature, and Culture' as the next step in the long road to conservation and sustainable economic development. It outlines seven defined strategies. The plan aims at the restoration of Lake Biwa and Yodo River Basin as a comprehensive, cultural, iconic and living entity.

Lessons learned

The recent history of Lake Biwa demonstrates how a major infrastructure development approach, designed primarily to meet the needs of downstream industrial and urban development, was transformed into a multi-stakeholder approach for the protection of the environment and biodiversity of Lake Biwa and its catchment.

Effective public involvement

Professor Kada Yukiko has been a leading actor in the long process, a campaigner for environmental justice and a founder of the Lake Biwa Museum, and in 2006 was elected Governor of Shiga Prefecture representing Japan's opposition Social Democratic Party. She speaks of Lake Biwa development as the 'obverse of development based on necessity', saying that it was not underdevelopment of water resources that failed to meet demands and caused social injustice, but overdevelopment of water resources that became the cause of social injustice. While tax revenues were used for long-term huge public infrastructure projects, the public was not consulted, leading to overdesigned interventions (some of which had to be scaled down in the face of economic recession), changed patterns of demand for water use, and the increasing priority given to environmental matters by stakeholder groups (Kada, 1999: 33).

Dialogue between water users, politicians, administrative authorities and water managers triggered change in tune with Japanese traditions, reflecting local priorities as well as regional and national preoccupations. This stakeholder participation, both from within 'the system' (academic, research and political groups) and from without (citizens' movements, protests and litigation), made a crucial contribution to achieving balance between the various uses of water and the needs of the resource itself.

The Lake Biwa experience highlights that the rights and entitlements of all populations relying on or benefiting from a water source have to be addressed and safeguarded with equal interest and vigour, if not with equal and similar measures. It has also shown the clear benefits of enhanced local sovereignty, local autonomy and local economy, once stakeholders were brought on board.

Balancing development and conservation

The Lake Biwa experience also shows that the physical management of water resources and associated infrastructure development can be done in a manner that also addresses environmental protection and ecosystem sustainability, and that water resources development and environmental sustainability can go hand in hand.

It shows how the interests and needs of upstream and downstream water-user groups can be reconciled through intensive engagement and processes of conflict resolution. Indeed, the conflicts provided opportunities to seek equity and synergy in water resources development as part of their resolution. The 'balancing' of development with conservation through conflict resolution and improved management in Lake Biwa thus saw

an 'integration' of industrial development, environmental sustainability, and the concerns and interests of groups in the upstream and downstream areas.

In this context, the Lake Biwa process also demonstrates how broader national development approaches impact upon water resource management, and how local approaches can influence national development.

Thus the impact of rice cultivation on the lake, both through the filling-in of wetlands and the diffuse pollution that drains from the fields, is in part a consequence of the Japanese policy of encouraging and protecting rice cultivation in areas where its viability would be questionable on strictly economic grounds.

The Lake Biwa case demonstrates both synergies and tensions between local sectoral development and lake management programmes and national efforts in the same domains. But management instruments emerged to address these as the need arose, which, together, were used to steer the complex programme of balanced development and conservation along a course leading towards sustainability.

The new approach began before the concept of IWRM was coherently formulated. However, while the term may not have been used explicitly in the literature of the past decades to describe the approach taken, the process followed is a practical example of how this philosophy of water resource management responds to practical pressures and real needs.

This progress was not achieved in a day or without mistakes. It required a long period of dialogue among different interest groups, ongoing negotiations to resolve contested issues, and the participation of all sections of the population.

Policies and institutions

While goodwill was generated through conflict resolution and dialogues, the outcomes would not have been sustainable without laws, regulations, appropriate institutions and clear management guidelines that can be monitored.

The Lake Biwa example suggests that a key factor in successfully institutionalizing innovative approaches is to start small, demonstrate results and then scale up. The approach taken to the problems of Lake Biwa, which were similar to those experienced in other parts of the country, provided a model for the national response – the local government tested appropriate laws that were successively scaled up to cover the country. The experience has been documented and developed to such an extent that it is now informing global action. In 1986 the Shiga Prefecture government founded the International Lake Environment Committee (ILEC), which has become a global lead organization and in 2003 launched the World Lake Vision.

A final lesson is that a conservation-led approach for developing IWRM processes and obtaining IWRM outcomes is indeed possible and can be successful, but constant vigilance, flexibility and continued actions are needed to sustain the outcomes.

References

IIC-JICA (Institute for International Cooperation, Japan International Cooperation Agency) (2005) *Japan's Experience in Public Health and Medical Systems*, Research Group, Institute for International Cooperation, Tokyo

Japan Water Agency, Lake Biwa Development Integrated Operation and Maintenance Office (2003) 'Lake Biwa Today', www.water.go.jp/kansai/biwako/english/development/d06e.html#a

Kada, Y. (1999) *Environmental Justice in Japan: Case Studies of Lake Biwa, Nagara River, Minamata and Niigate-Minamata*, paper presented at a workshop on Public Philosophy, Environment and Social Justice held in New York, conducted by the Carnegie Council on Ethics and International Affairs, New York

Kira, T., Ide, S., Fukada, F. and Nakamura, M. (2005) *Lake Biwa – Experience and Lessons Learned Brief*, Kusatsu, Japan: The International Environment Committee

Lake Biwa Museum (2008) Lake Biwa Facts, (accessed at) http://www.lbm.go.jp/english/facts/

Masaki, Takakura (2008) 'Global Warming Imperils Rare Fish in Lake Biwa', *The Yomiuri Shimbun*, 4 March

Nakamura, M. (1998) 'Regional Environmental Planning: The Lake Biwa Development Project', in Cruz, W., Takemoto, K. and Warford, J. (eds) *Urban and Industrial Management in Developing Countries: Lessons from Japanese Experience*, Economic Development Institute of the World Bank, Washington

Nakamura, M. (2007) *Lake Biwa: Management and Research*, Malaysian National Colloquium on Lakes & Reservoir Management, Status and Issues, 2–3 August 2007 at National Hydraulic Research Institute of Malaysia

Shiga Prefecture (2008) *Environment of Shiga*, (accessed at) www.pref.shiga.jp/multilingual/english/environment/files/4syo.pdf

Ueda, T., Kawabata, A., Koitabashi, T. and Narita, T. (1998) *Data of Regular Limnological Survey of Lake Biwa*, Center for Ecological Research, Kyoto University Technical Report No.1

10

Taming the Yangtze River by Enforcing Infrastructure Development under IWRM

Yang Xiaoliu[1] and Mike Muller

China's Yangtze is the third longest river in the world and also third largest in terms of annual flows. It is deeply rooted in China's history and culture. Over the centuries, the Yangtze's floods have killed tens of thousands of people, destroyed millions of homes and caused incalculable economic damage. On the positive side, the river is a source of water for people and their agriculture and a potential source of clean electricity to power the burgeoning economy, and has long been an important transport artery.

While China's approach to water infrastructure – as exemplified by the massive Three Gorges Dam and related projects on the river – has been heavily criticized, the country has gone to great lengths to ensure that social equity and environmental sustainability are also respected.

The Yangtze demonstrates that an integrated approach to infrastructure construction and management arrangements can support economic growth and make a substantial contribution to the achievement of more equitable regional development, while still protecting important ecological functions of the river.

The development context and water challenges,

With a total length over 6300 kilometres, the Yangtze is the third longest river in the world, shorter only than the Nile and the Amazon. Flowing eastward to the Pacific, it has an unrivalled geographic reach, spanning across West, Central and East China (see Figure 10.1).

The river's basin is rich in water resources. Its yearly available water amounts to 996 billion cubic metres, accounting for 36.5 per cent of China's total. The Yangtze River also has abundant hydropower resources, with a current annual power output of an estimated 1.19 trillion kilowatt hours, 49 per cent of the nation's total hydropower output. The basin is the focus of China's hydropower development, with a technically exploitable potential of 256.3 gigawatts, less than one-third of which has been developed. It accounts for 48 per cent of the country's hydropower potential, and when economic viability is factored in the figure goes up to 60 per cent (General Institute of Water Resources and Hydropower Planning and Design, 1994).

The Yangtze Basin is also rich in biodiversity, accounting for two-thirds of China's 3,980 genera and half of China's nearly 30,000 spermophytes. The river itself is home to more than 370 fish species, nine of which are on the list of rare species under high protection by the State. The basin is thus also China's treasure house of rare and precious aquatic wildlife, justifying its international importance in biodiversity and wetland protection. None of the other rivers in China is comparable to it in terms of ecological functions.

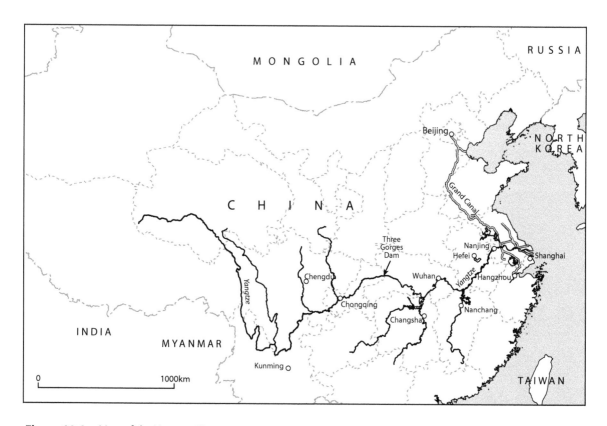

Figure 10.1   Map of the Yangtze River

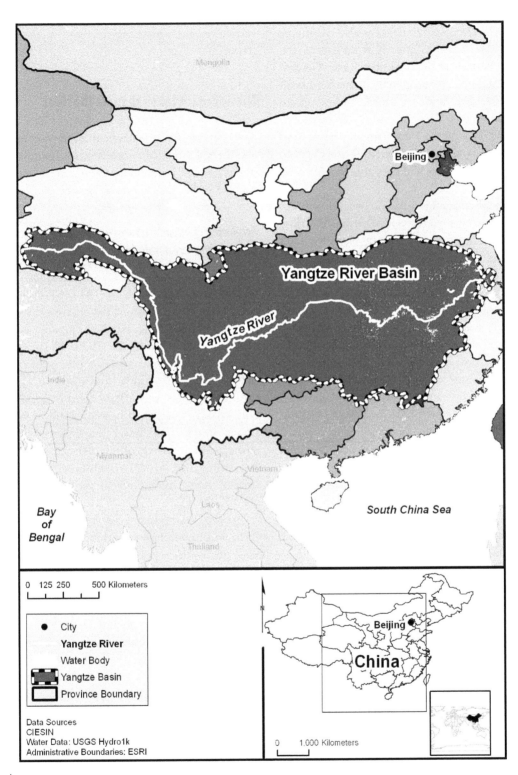

Location map

However, the Yangtze River is also the major inland waterway in China; it is known as the 'Golden Waterway' connecting East, Central and West China. Its navigable channels on both the mainstream and tributaries adds up to 57,000 kilometres, 52.5 per cent of China's total. The 2,837-kilometre mainstream navigation channel has a shipping capacity equivalent to four to six railways, each of the same length as the river.

The water resources in the Yangtze River Basin are huge in volume but unevenly distributed in time and space. The population of the basin has long faced dangers from flood disasters. Floods are, however, only one of the problems.

The river flow during flood season is hardly utilized though it accounts for 70–75 per cent of total yearly water, while in the dry season water is short and the river has a low navigable depth. There are also challenges of drought, water pollution, and water and soil erosion. Some parts of the basin suffer from a groundwater overdraft while, at the mouth, there are problems caused by saline intrusion into the river's estuary.

Given all these factors, a great deal of effort has gone into reducing the threat of flooding, including, for example, the construction of the Three Gorges Project, as well as numerous other works including flood detention areas, expanded flood protection dykes and storage, power generation and navigation infrastructure.

However, intervention brings its own challenges and the basin is also facing the issues of reservoir resettlement, compensation to the people due to flood detention, flood insurance and ecological stress.

Drought and water and soil erosion are becoming increasingly serious in the upper and middle reaches of the river, and serious conflict between water supply and demand occurred in some upstream areas. There is concern (e.g. Cai, 2005) that ongoing economic and social development activities such as city expansion, road construction, energy exploitation and mineral extraction are worsening the state of the water resource.

The approach taken to tame the tiger

The Yangtze Basin is larger than most countries, and its management encompasses many of the dimensions of national development. However, a number of key thrusts can be identified.

Accelerating infrastructure construction for basin and national economic development

The first hydropower station on the Yangtze River was built in 1910 and, by 1949, 31 dams had been built in the whole valley; these were small, however, with a total installed capacity of only 134 megawatts. From the 1950s to the 1960s, hydropower stations and reservoirs were rapidly developed in the middle and lower reaches of the river; from the 1960s to the 1980s, the construction of hydropower stations in the upper, middle and lower reaches were in full swing; while at the end of the 20th century and the beginning of the 21st century, hydropower development was mainly in the upper Yangtze (MWR, 2001).

To date, more than 2,440 hydropower stations have been built, with a total installed capacity of 69.7 gigawatts, representing 70 per cent of the nation's total hydropower capacity (CWRC, 2004). In addition, more than 44,000 reservoirs have been constructed with total storage capacity of 137.3 billion cubic metres, including 109 large-scale reservoirs (over 100 million cubic metres) with a total storage capacity of 66.7 billion cubic metres, and 997 medium-sized reservoirs (between 10 and 100 million cubic metres) with a total storage capacity of 24.2 billion cubic metres. These reservoirs, together with their associated infrastructure of canals, pipelines and pumps, make up a water supply network that can provide a huge amount of water to support people's social and economic development. In 2004, 189.9 billion cubic metres were supplied to different users (CWRC, 2005).

Enforcing the concept of IWRM to coordinate development and protection

There has been a growing recognition that water management in China requires a resource-oriented focus and not just infrastructure projects (see section 'from dams to laws', p140, below). In recent years, the existing basin institutions and related administration departments at all levels have been required by the central government to maintain the health of the Yangtze River (Huang, 2007). It has been emphasized that protection of the environment and the construction of infrastructure should receive equal attention (Wang, 2002). The government also has encouraged people to save water consciously so as to achieve water conservancy and build a conservation-minded society (Wen, 2004).

The objective is to manage water resources based upon the carrying capacities of water and environment at basin level. In this context, the need for the distribution of productive forces and industrial activities to take into account water availability and environmental protection in the basin has been emphasized – with productivity levels determined by the environmental and ecological situation. Concretely, five dimensions are addressed in planning:

- urban and rural areas;
- city and region;
- society and economy;
- people and nature and
- basin and nation.

Economic development is seen as the foundation for the coexistence of people and nature. The basic principle of 'ensuring the health of the Yangtze to promote harmony between humans and nature' guides water resource management in the basin and has led to reforms in the water pricing system, water pollution controls, safeguards for drinking water security and infrastructure planning (CWRC, 2001 and 2005). In these reforms, five key tasks were addressed.

1 Pollution control

With the objective of improving the water body's functions, pollution control has been intensified, particularly in the key areas along the Yangtze such as the Three Gorges Reservoir, the Danjiangkou Reservoir (the source for the South-to-North Water Transfer Project) and all schistosomiasis impact areas (schistosomiasis is transmitted when wetlands with snail vectors are exposed to sewage). At the same time, extensive sewage treatment facilities have been constructed. For some key places, such as source areas for drinking water, special protection rules have been worked out and strict protection measures adopted.

Water pollution monitoring systems have been established to oversee waste disposal. Water function zoning for various purposes has been carried out to determine the pollutant-bearing ability of each zone. Important water bodies are monitored in real time with early warning mechanisms to deal with pollution accidents. Laws and regulations governing water pollution control have been promulgated at different levels. A long-lasting stakeholders' participation mechanism for joint work was set up for collaboration and cooperation among the line departments concerned, including basin authorities, and local governments relevant to water conservancy, environmental protection, urban construction and land reclamation.

2 Water and soil erosion control

Integrated planning, monitoring and research on water and soil conservation have been reviewed and new approaches implemented. Comprehensive management is used to control human-induced erosion, guided by laws promulgated by the central government and regulations established by basin and local authorities.

3   Flood control and disaster reduction

Special attention has been given to the rehabilitation and construction of infrastructure – such as reservoirs, flood diversion channels, river adjustments and embankment reinforcement – that greatly increase the basin's flood control capacity. A number of defective reservoirs have been reinforced. A flood management system has been established that comprehensively regulates flood control protection areas, flood storage and detention areas, floodplains and planned reserve areas. Based on these, a holistic management system able to deal with risk evaluation, social and environmental impact assessment, and compensation due to flood detention, is under development. A flood insurance system is also being developed. In addition, effective means to use floodwater as a resource are also being studied.

4   Integrated utilization of water resources

A wide range of infrastructure has been developed to ensure water supply for cities and key industrial areas, and to secure drinking water for residents as well as to balance supply and demand for water in areas where drought and water shortage prevail. Irrigation systems are continuously being constructed and improved. Construction of the backbone hydropower projects in the upper Yangtze was accelerated to support national development, while small hydropower stations for rural electrification have also been built. Regulation of navigation routes has been intensified, while construction of ports and piers has been undertaken to improve traffic capacity and efficiency. In some cases, these developments competed with each other, requiring regulatory intervention and guidance; in others, they offered opportunities for multi-purpose designs.

5   Solving the ecological and environmental problems caused by the water resources and hydropower infrastructure

Development of water resources and hydropower infrastructure can bring huge economic benefits, but it can also have negative impacts on society,

ecosystems and the environment, and these tradeoffs need to be properly addressed. The major social issue in regards to the Yangtze is resettlement due to reservoir construction. A long-term systematic method is being used that considers factors such as environmental carrying capacity, financial support, self-reliance and long-term benefits relevant to the people resettled.

Sedimentation is a key environmental issue. Prototypes were used for measurement and real-time monitoring was enhanced. To gain approval from the central government, any proposed infrastructure project in the Yangtze Basin must demonstrate that it addresses flood control, power generation, navigation and sediment regulation in a holistic way (MWR, 2001). Impacts on aquatic biota are also taken into consideration when any infrastructure is developed; basic environmental flows and fishways are now obligatory. Where necessary, artificial breeding of affected fish species is required to minimize the effect on aquatic biota.

Giving overall consideration to wetland protection, biodiversity protection, navigation security, water supply, power generation and flood control, a special plan for managing water intakes was worked out and implemented, using biological and ecological engineering approaches for interventions such as man-made wetlands.

From dams to laws – the instruments used to achieve harmony

Few major water projects have been studied in as much depth and come under such intense external scrutiny as the Three Gorges Dam. Yet it is not generally recognized that the project has been a catalyst, driving the introduction of new approaches to water management in China.

The idea of building a dam on the Yangtze to generate power and control the devastating floods has a long history. The nationalist leader Sun Yat Sen proposed it in 1919 as a key element of an industrial plan for a re-emerging China. The idea gained currency and was supported by Sun Yat Sen's American allies, who offered their own experience as a model for the development of the Yangtze. Anticipating the current environmental controversies by 60 years, US academic and student of China John Fairbanks noted:

> The one American achievement which has most appealed to Chinese observers as an American model for China to follow is the regional development program of the Tennessee Valley Authority. TVA makes sense in China ... The fact that in our own more fully developed economy we have less urgent need, or think that we have less urgent need, for such programs of regional development, should not prevent our using the TVA idea in our foreign policy.
>
> (Fairbanks, 1948)

However, it was almost 50 years later before the vision of the Three Gorges Dam was translated into a reality. But even as construction of the dam got under way in the early 1990s, and in part because of the controversies that arose about the dam, the approach to water resource management was coming under review. In March 1999, the then-new Minister of Water Resources Wang Sucheng, in an address to the Chinese Hydraulic Engineering Society, called for a new perspective and set out a crisp statement of the need for IWRM, although he called it 'resource-oriented water management':

> When the society and economy continued to develop, people made growing use of water until some day they suddenly found that there was insufficient water. Moreover, the problem of water shortage becomes more and more serious and obvious worldwide. At present, water shortage severely constrains further social and economic development in China, especially in North China. Pollution as a result of rapid industrial development also came to the surface, worsening the shortage of water resources. Given the situation as such, water saving and preservation of water resources have become our increasingly important task and will assume tantamount importance in the future. Traditional project-based water management often focused on single projects e.g. planning of a river was mainly about the number of cascades to be built rather than considering the relationship between water resources and economic development. In the planning of the Yangtze River basin, great attention was paid to the role of the Three Gorges while the Yellow River basin planning emphasised the number of cascades. While this kind of work was correct in view of the prevailing productivity level at the time, they can no longer cater to the continuous social and economic development under the current circumstances.
>
> (Wang, 2002)

The response has been the much stronger focus on the resource-oriented water management that this case study describes. The major challenge now is to proceed with the institutional reforms needed (see Varley, 2002) to enable a resource-oriented approach to work.

Strengthening integrated basin management

A key element in China's resource-oriented approach has been the coordination between basin management and district management at the levels of province, city and county. This has to be carefully handled in aspects of jurisdiction, accountability and institutions by way of consulting mechanisms and information exchange. For each involved body the management is integrated and the role is clearly defined as indirect management, dynamic management and post-supervision. Through this approach, an effective and harmonious mechanism of basin management is being operationalized in the basin. A transdepartmental and transprovincial management institution (Cheng, 2007) was established, along with extensive participation mechanisms for all parties concerned, so that all natural resources, including water, soil and biota, could be well protected and reasonably utilized, while economic and societal development could be sustained.

Strengthening the legislative setting

The legal system has been further enhanced at all three levels: law, administrative regulation and departmental sub-regulation. Legal documents such as 'Regulations for the Water Resources Protection of Yangtze River Basin' and 'Management Measures for the Yangtze Estuary' have been enacted, and a comprehensive legal system is being built up.

Water-related laws such as the Water Law of PRC (NPC, 2002), Flood Control Law of PRC (NPC, 1998), Soil and Water Conservation Law of PRC (NPC, 1991) and Water Pollution Control Law of PRC (NPC, 1984) were worked out and provided a sound basis for the holistic management of flood, water resources, land reclamation, water and soil conservation, and water pollution. In the case of the Yangtze, other factors such as sand excavation and groundwater tapping are also integrated into the management.

Revising plans in view of IWRM

Planning is the key to ensure a healthy Yangtze River. The 'Comprehensive Utilization Planning of Yangtze River Basin' has been revised from an IWRM perspective, in which water infrastructure is considered in the context of overall resource management, and regulation and protection are highlighted more than ever before.

Promoting public participation and democratic practice

The idea of a healthy Yangtze has been widely promoted to the public so that people can be mobilized to support activities to protect the Yangtze. Information-sharing systems have been set up amongst stakeholders to establish a democratic course for governments' decision-making and assert people's right to know and take oversight of governments' work. Through various activities such as workshops, seminars and forums (e.g. the Yangtze River Forum), views on important subjects are exchanged and discussed among stakeholders. Ideas are collected and analysed to seek agreement regarding the protection and development of the basin.

Water resources management in a harmonious society – the three Es

The development strategy of the Chinese central government is to 'put people first, fully implement the Scientific Outlook on Development and accelerate the building of a harmonious socialist society' (Hu, 2003). One of the basic and strategic steps for the whole country is to achieve harmony between man and water.

In some upstream areas of the Yangtze, water resources were unexploited due to difficult terrain, weak economies and limited technology. In the more distant past, land and forests were overused for economic reasons with the result that water source areas lost protection. As a result there were sharp contradictions between people and water: flood and drought disasters frequently occurred, land was gravely eroded, and water quality quickly deteriorated.

The implementation of the development strategy has meant that all three Es were given priority – economic efficiency, social equity and environmental sustainability. Great efforts were made to conserve water and soil by enforcing both water resource exploitation and water saving, balancing economic development and social equity, and coordinating development and protection. In this context, large-scale infrastructures were provided for water transfer, water supply, river harnessing, flood control, drought relief, and especially drinking water safety. At the same time, to help maintain a balance between urban and rural development, considerable attention was paid to irrigation systems, to enhance agricultural productivity, as well as to small-scale water conservancy projects and hydropower plants, with accessible drinking water given a high priority

(CWRC, 2005). In terms of conservancy and protection, 'integration' was set as a goal, the relevant national policy of water resource conservancy and protection was comprehensively implemented, and water conserving was taken as a long-term strategic guideline.

A three Es river needs common efforts

The objective of building a harmonious society has meant that the whole Yangtze River Basin has been taken into consideration. The need for development balance between city and village were emphasized, between locality and region and between society and economy, through comprehensive development planning and integrated management. To implement the new Water Law (NPC, 2002), a mechanism was set up to take integrated measures for economic development and environmental protection and coordinate the relationship between upstream and downstream.

In this context, the people living along the upper Yangtze have made huge contributions by preventing soil erosion, expanding forestation and conserving water and soil. In these efforts, they were supported financially and technically by the middle and lower Yangtze as well as the central government. In order to support and help upstream areas develop their economies and protect the environment, supportive and compensation policies were prioritized to realize protection while developing, with the expectation that a Yangtze satisfying the three Es appears. The concept of a 'green GDP' is gradually being accepted by governments and the principle of 'polluter pays' is popularly recognized (CWRC, 2005).

In the downstream area, the integrated Yangtze flood control system has already helped to secure the lives and properties of local residents. This included opening dyked areas to create space for flood flows and giving up farmland to be returned to lake. Water resources protection and water pollution prevention have been continuously enhanced, and substantial progress was made in various aspects of ecological reconstruction and environmental protection (Cai, 2005).

The road to integration and harmony

In realizing IWRM, the Scientific Growth Outlook has been adopted and implemented; traditional ideas relying only on infrastructure were transformed into an integrated approach that actively sought to secure the health of the Yangtze. Reflecting this change in thinking, 'three principles' have been followed to coordinate 'three relationships' by emphasizing four aspects of work.

'Three principles'

1 Harmonizing people and water

The harmony between people and water is not only a fundamental requirement for sustainable growth and all-round development of the people, but also a concrete reflection of the construction of a harmonious society. Harnessing the Yangtze River lays foundations for developing the basin to further accelerate economic and social growth. Whether harnessing and protecting or developing and utilizing the Yangtze River, the ultimate goal is to serve sustainable growth and all-round development of the people, which requires that the significance of harmony between the people and water is recognized and respected.

2 Respecting nature

Throughout history, the appearance, growth and disappearance of rivers has been a part of natural ecological cycles, but these cycles have also been impacted by human activities. If humans develop resources such as rivers, shorelines and wetlands like a plunderer, the pace of the disappearance of rivers will be quickened, which will result in water shortages and deterioration of the ecological environment, and will punish humans by constraining their survival and growth. Only if rivers are developed and protected systematically can the ecological system of rivers achieve a favourable cycle and be utilized by humans. It must be considered whether or not harnessing and developing the Yangtze River can do harm to its essential characteristics and how its health can be safeguarded.

3 Prioritizing economic development

Human societies are based on economies, which effectively coordinate the interaction between people and nature. In recent years, as the economy grew increasingly fast, pressures on the water resources and the environment in the Yangtze River Basin also grew. Economic growth depends on the consumption of resources, but it was noted that the consumption of water resources per product was higher than the world average. Similarly, wastewater drainage per unit GDP is several times that of developed countries. To remedy this requires that the carrying capacity of water resources and the environment be considered, and that efforts be made to base economic growth on the efficient utilization of resources and the reduction of environmental pollution.

'Three relationships' to coordinate

1 Relationship between city development and infrastructure construction

On the one hand, infrastructure to control floods is important to create favourable conditions for the development of cities and industries. On the other hand, urban development must be undertaken in a manner that dos not create flood risks. This needs close cooperation between the water sector and the urban development sector, which are normally separated in governmental organization in China.

2 Relationship between consumption and conservation

High water-consumption industries such as power generation, chemistry and metallurgy are encouraged to locate themselves near the river for ease of water supply. New industries that are high-tech, have low water consumption and create less pollution are allowed to locate anywhere in the basin.

3 Relationship between utilization and protection

Ecological water demand has to be met and minimum flows maintained to prevent seawater intrusion and sedimentation, to keep waterways clear and to maintain the natural ecological system. Attention has to be paid to the ecological balance for the protection of wetlands, rare animals and plants.

Four aspects of the work

The first emphasis has been to improve planning. It is necessary to update comprehensive plans for the water resources of the Yangtze Basin within the framework of the national water resources strategic plan and the general layout of the south–north water transfer project. The comprehensive plans, regional plans and speciality plans for each branch river can be revised – especially the hydropower plan for main and branch rivers – on the basis of the comprehensive water resources plans. In addition, the relationships between the upper, middle and lower streams have to be addressed, as well as the balance between economic development and environment protection, the local and the whole, the immediate and long-term benefits, and so on. The approach to planning is being adjusted to give priority to the protection of water resources – harnessing water resources reasonably while protecting them, and protecting while developing. The ideal result is a utilization of water resources that achieves win–win outcomes in water conservancy, hydropower, waterway exploitation and basin environment, as well as ecosystem protection.

The second emphasis has been to perfect the establishment of water laws and codes and strengthen their implementation. These include the following laws, provisions and procedures: the New

Water Law, the Flood Control Law, the Water and Soil Conservation Law, the Law of Water Pollution Protection, the Environmental Protection Law, and the Environmental Assessment Law. While these actions strengthen administration of water resources and its execution, more is still needed. A Yangtze River Law and Regulation of Water Resources Protection on the Yangtze Basin are under preparation and should be issued soon. Regulation of Multiple Purpose Reservoirs is also needed to share the investment, as well as to distribute the benefits, of serial reservoirs and hydropower stations. Problems such as the compensation of environment and ecosystem services at upper and lower streams still need to be further addressed, especially the regulations on relocatees from large- and medium-scale reservoir construction (Fan et al, 2005). These should aim to create conditions for reservoir relocatees to live and work in peace and contentment, cast off poverty, and become better off.

Water projects also need to follow the regulations on comprehensive planning of basin water resources, and obtain planning authorizations, water licensing permission for power development and environmental assessments. In operations, stated flood storage capacities established in terms of the Flood Control Law need to be observed so as to achieve reasonable flood control levels. Already, to manage both floods and droughts, unified releases and management are guided by the authority in charge of flood prevention and drought mitigation, with the costs in terms of electricity production carried by the owners.

The third emphasis is to complete the mechanisms of basin management. The Changjiang Water Resources Commission (CWRC) is still undergoing a step-by-step reform of its management systems to become a proper basin commission (Cai, 2005). It is intended that it will be composed of representatives from the Ministry of Water Resources, the State Development Reform Commission and its related departments and bureaus, as well as representatives from provinces and municipalities and water users in water conservancy, hydropower, navigation and water supply, including public representatives, specialists and civil society organizations. With leadership from the Ministry of Water Resources and the State Development and Reform Commission, this trajectory should see the CWRC becoming an administrative institution for sharing information and overseeing water uses – an institution that is guided by the macro-level direction of government and market principles, that participates in public discussion and consults democratically, and eventually that establishes a new mechanism for integrated management of the basin under the law.

The fourth emphasis is for different administrative units to carry out their extensive operational duties faithfully. CWRC's duty is to maintain the water environment and water ecosystem on the Yangtze River, coordinating matters such as the relationship between up- and downstream, right and left shores, different provinces, as well as water quantity and quality, and so on. According to the Water Law (NPC, 2002), CWRC must also integrate the releases, operation, management and supervision of water quantity and quality at important reservoirs, hydropower stations and water transfer projects, thus safeguarding the environment as well as flood control and storage against drought. This work must be coordinated with that of the Yangtze River Water Resource Protection Bureau, set up in 1983 by the State Environment Protection Bureau and the Ministry of Water Resources, and with other basin management administrations such as the lead upstream water and soil conservation group (established in 1989) and the commander of Yangtze Basin flood control and drought mitigation.

The water 'functional areas' and the total pollution absorption capacity at river-reach level still need to be determined to guide the adjustment and optimization of the layout of water catchments and drainage for cities along the river. Another major focus will be the establishment of the Danjiangkou reservoir (the middle route project of the south–

north water transfer project), Sanjiangying (the east route of south–north water transfer project), and the water conservation district at the mouth of the Yangtze River.

The outcomes

Given the huge challenges and the comprehensive scope of the reforms that have been undertaken, it is not easy to document their outcomes, which can only be the first steps of a long journey. However, some elements, such as infrastructure, are visible. Thus more than 45,000 reservoirs have been constructed with a total storage capacity of 148 billion cubic metres (not including the Three Gorges reservoir), as well as 2,441 hydropower stations with a total installed capacity of about 0.7 billion kilowatts. The water supply capacity has reached about 200 billion cubic metres. Over 30,000 kilometres of dykes have been enhanced, and an effective irrigated area of 150 million hectares supplied. Thousands of flood control reservoirs, flood retention zones and flood diversion channels have been built together with non-structural measures.

While the expansion of hydropower has been controversial, one result has been a significant reduction in the expected levels of China's national carbon emissions (see Figure 10.2).

It is also possible to document outcomes from the systems that were put in place for flood forecasting, warning and regulation for mainstreams and tributaries, to protect the whole basin from disastrous floods – these have been tested. These interventions are credited by external observers with preventing major disasters. Comparing the Yangtze floods of 1998 to those in Zeeland in the Netherlands in 1953 and on the USA's Mississippi in 1993, the Rand Corporation concluded that:

… the lessons of history are that, while determining safety levels might be defensible on cost-benefit or IWRM bases, the planning for regional infrastructure and services must cover total catastrophic breakdown and must include secondary, contingency responses that can be invoked when primary responses are overwhelmed. In Zeeland, lack of such planning led to catastrophe, but in the Yangtze case, this planning was a major reason why loss was only a fraction of what had been suffered in previous floods.

(Kahan et al, 2006)

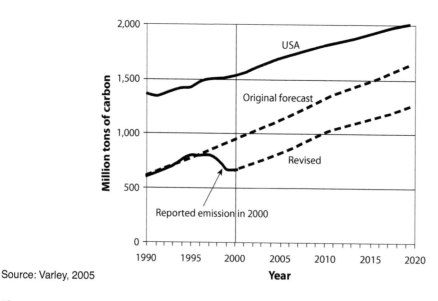

Source: Varley, 2005

Figure 10.2   USA and China – carbon emissions

Substantial attention has more recently been focused on the social and environmental issues in large projects, which have become increasingly controversial. Although external reviews again suggest that China's performance has been comparatively good – a World Bank study of involuntary resettlement praised China's improvement and performance in almost all respects (Varley, 2005) – the costs of addressing these issues is substantial and must be provided for. As an example, it is reported that the cost of the first phase of the eastern and middle routes of the south–north water diversion has risen from 101 billion Yuan to a total of 225 billion Yuan (almost US$28 billion), largely to meet environmental protection and relocation costs (Xinhua, 2006a).

Turning to economic dimensions again, navigation is critical. The Yangtze's navigable waterways amounted to 57,000 kilometres, about 52.5 per cent of the national total. While dam construction has reduced the navigable length available at national level, there have been recent efforts to reverse this. One project, supported by the World Bank, which included tributaries of the Yangtze, targeted regional development priorities:

> The project … was intended to stem the trend towards a diminishing waterway network by expanding capacity and modernizing the waterways of Zhejiang, Hunan, and Guangxi provinces. These provinces were selected for the project because of their potential for economic growth, the expectations of traffic increases along their inland waterways, and the year-round navigability of the waterways concerned. Guangxi has the highest incidence of poverty among the three provinces with nearly 35 percent of its counties designated as 'poverty counties'. … Through this mix of provinces, the project contributed to the government's objective of reducing the development imbalance between the fast-growing coastal provinces and the less prosperous inland regions.
> (World Bank, 2005)

The economic importance of water management for navigation is highlighted by a recent claim by China's Ministry of Communications that the Yangtze is now the world's busiest freshwater route. In 2006, the Yangtze's annual freight volume of 640 million tonnes was 1.6 times that of the Mississippi and 2.3 times that of the Rhine (Xinhua, 2006b).

Pollution control efforts have also had significant impact. An external review cited evidence suggesting that the rate of waste water treatment at national level increased from 60 per cent to more than 80 per cent in the period 1990–1998 (Varley, 2005) (see Figure 10.3), and the trend has continued since then. Indeed, more recent commentary suggests that there may have been over-investment in sewage treatment. One independent study conducted under Canadian Aid for the Chinese Ministry of Construction, which included several plants in the Yangtze Basin, found that some of the plants were operating significantly under their design capacity and that therefore 'a large portion of the capital investment is not working to reduce pollution' (Palmer et al, 2000). This demonstrates that, while there has been significant investment in wastewater treatment, it will in future have to be matched by investments in the wastewater collection infrastructure.

To protect the aquatic environment and fish stocks, fishing was banned in many reaches of the Yangtze. Natural protection areas include: the fish

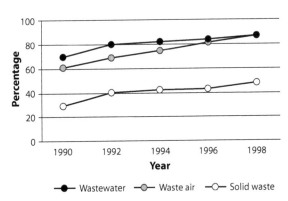

Note: Data for CAOEs only. TVIEs have substantially lower treatment rates although they too have been improving. *Air, Land, Water* 2001. Source: Guo (1999).

Source: Varley, 2005

Figure 10.3   Industrial waste treatment rates, 1990–1998

spawning enclosure at the estuary of the Yangtze River; the Chinese sturgeon protected zone in Zhicheng; Yichang city below the Gezhouba Dam; and the white-flags dolphin and cowfish natural protection area, which stretches from Luoshan reach at Shisou in Hubei Province, through Hukou reaches in Jiangxi Province to Tongling reach.

It is perhaps in the institutional arena where the most remains to be done. As has been highlighted, the Changjiang Water Resources Commission (CWRC) does not yet have the powers of a fully fledged Basin Commission; one consequence of this is a dispersal of powers amongst decentralized local administrative structures. According to a World Bank evaluation:

> The decentralization of political and economic power, to separate riparian provinces on the great river systems, has made central government control over allocation of water resources more difficult. Many water management decisions are 'upstream-downstream' issues cutting across administrative boundaries e.g. the top-end capture of river flows, waste disposal downstream by urban polluters, how to share the costs and benefits of expensive hydraulic infrastructure between provinces. The existing water management framework is a result of history and is now insufficiently comprehensive or unified to optimize uses and resolve conflicts in a socially and economically optimal manner. Negotiation is the key discipline for mediating between sector and regional interests (e.g. industry and agriculture, upstream-downstream), and between the ministries responsible for different aspects of WRM and planning. There are multiple political jurisdictions and the provinces compete with each other. Central government has played a major technical and leadership role especially when large projects pose difficult engineering problems and beneficiaries are distributed between multiple riparian provinces. A river basin is the most appropriate unit for negotiation and water resources management. This requires institutions corresponding to basins, rather than provinces, and these have proved very difficult to establish, despite much central government commitment and awareness.
>
> (Varley, 2002)

The World Bank report (2005) further points out several other institutional challenges for water resources management in China:

- The Ministry of Water Resources (MWR), while it has been effective at supply-side interventions, has limited ability/authority to carry out demand-side interventions.
- While water management agencies have the necessary technical skills to carry out their mission, appropriate internal incentives are weak or missing.
- The provincial water resources departments, currently the de facto authority for water resources management (and to a large extent development), are strongly influenced by specific interests and thus unable to take a holistic approach.
- 'The MWR-controlled river-basin commissions (RBC) continue to operate with large staffs, but with declining authority and reduced ability to manage integrated water resources development. Incremental progress in designing institutions for river basin management has been made, but limited implementation achieved' (Varley, 2002).

Conclusions and key lessons learned

An integrated approach to infrastructure construction and management can make a substantial contribution to the achievement of more equitable regional development up- and downstream. If such development is combined with a clear appreciation of the need to build harmony between people and nature, sound outcomes can be achieved. From the experience in the Yangtze, some concluding remarks can be made, as follows:

1   From the development of the Yangtze River Basin, it is recognized that IWRM can help harmonize development and protection of a basin, which are always thought to be in

contradiction of each other. The practice along the Yangtze shows how IWRM can be promoted through infrastructure development and how infrastructure development can be realized in a harmonious way, with equal attention to economic development, environmental protection and social equality.

2 Strengthening IWRM needs solid technical support. One element of this is to qualitatively and quantitatively define the environmental flows, which are essential for planning, design, construction and management of water infrastructure in the context of IWRM. Environmental flows must be worked out from the perspective of the overall basin and its characteristics.

3 Ecological and environmental monitoring systems have to be set up and regularly operated in the whole basin to provide sound data as a basis for harmonious basin development planning, and the establishment of an operational ecological and environmental compensation mechanism.

4 In a basin as large, complex and dynamic as the Yangtze, water resource management must be part of a national development vision to be successful. While deepening the various reforms of production-sector relationships and promoting the growth of productivity, a harmonious society between people can be constructed that can also ensure the coexistence between people and nature. Only joint efforts for the protection and construction of a healthy Yangtze River can succeed!

Note

1. Yang Xiaoliu is a Professor of Water Resources at Peking University, China. He worked for nine years with China's central government.

References

Cai, Q.H. (Commissioner of Changjiang Water Resources Commission) (2005) *Ensure the Healthy Yangtze and Promote the Harmony between Human and Nature*, keynote speech at the First Yangtze Forum, Wuhan

Cheng, W.G. (Vice-governor of Sichuan Province) (2007) *Build up Scientific Outlook of Development, Endeavor to Build the Ecological Barrier of the Upper Yangtze River and Make Contribution to the Prosperity of the Yangtze River*, available online at http://eng.cjw.gov.cn/detail/20070118/80010.asp

CWRC (Yangtze (Changjiang) Water Resources Commission) (1998) *Atlas of the Yangtze River Basin*, China Map Press, Beijing

CWRC (Yangtze (Changjiang) Water Resources Commission) (2001) *Comprehensive Planning for the Yangtze River*, technical report

CWRC (Yangtze (Changjiang) Water Resources Commission) (2004) *Re-examination Results of Water Power Resources in China. The Yangtze River Valley – Introduction.*

CWRC (Yangtze (Changjiang) Water Resources Commission) (2005) *Yangtze Declaration on Protection and Development*, available online at www.cjw.gov.cn/ad/yangtzeforum/detail/20060623/63157.asp

CWRC (Yangtze (Changjiang) Water Resources Commission) (2007) *General Work Plan for Revising the Comprehensive Planning for the Yangtze River Basin*, internal working paper

Fairbanks, J. (1948) *The United States and China*, 1st edition, Harvard University Press, Boston

Fan Jihui, Cheng Genwei, Zhangyan, et al (2005) *Existing Problems and Suggestions for Hydroelectric Cascade Development in Upper Minjiang River*, China Water Resources Journal, October

General Institute of Water Resources and Hydropower Planning and Design (1994) *Atlas of Planning for Large and Medium-scale Power Station in China*, Ministry of Water Resources, Beijing

Hu, J. (Chairman of the People's Republic of China) (2003) Speech delivered at the 6th Plenum of the 16th Communist Party of China (CPC) Central Committee, Beijing

Huang, L.X. (Vice-governor of Jiangsu Province) (2007) *Ensuring the Healthy Yangtze and Securing Sustainable Socio-*

Economic Growth of Jiangsu Province, available online at http://cjw.gov.cn/detail/20070115/79733.asp

Kahan, J.P., Wu, M., Hajiamiri, S. and Knopman, D. (2006) *From Flood Control to Integrated Water Resource Management: Lessons for the Gulf Coast from Flooding in Other Places in the Last Sixty Years*, Rand Corporation Occasional Paper, Santa Monica, CA

MWR (Ministry of Water Resources) (2001) *Outline of the Yangtze River Planning*, China Water and Power Press, Beijing

NPC (National People's Congress) (1984) *Water Pollution Control Law of PRC*, Beijing

NPC (National People's Congress) (1991) *Soil and Water Conservation Law of PRC*, Beijing

NPC (National People's Congress) (1998) *Flood Control Law of the People's Republic of China*, Beijing

NPC (National People's Congress) (2002) *Water Law of the People's Republic of China*, Beijing

Palmer, M, Lanqing Jia, Zhang Yue, Wang Lin and Fritz, J. (2000) *Preliminary Assessment of Municipal Wastewater Treatment Plant Operations and Biosolids Management in China*, available online at www.h2o-china.com/lianmeng/21cnwater/eng-art/2001/mp.htm

Varley, R.C.G. (2002) *The World Bank's Assistance for Water Resources Management in China*, World Bank Operations Evaluation Department, background paper for China Country Assistance Evaluation, Washington, DC

Varley, R.C.G. (2005) *The World Bank and China's Environment 1993–2003*, The World Bank Operations Evaluation Department, Washington, DC

Wang, S. (2002) *Resource Oriented Water Management: Towards Harmonious Coexistence between Water and Nature*, China Waterpower Press, Beijing

Wen Jiabo (Premier of the Government of China) (2004) *The Government Work Report 2004*, Xinhua News Agency, Beijing

World Bank (2005) *Project Performance Assessment Report, Report No.: 34595, People's Republic of China Inland Waterways (Loan 3910-CHA)*, 5 December, World Bank, Washington, DC

Xinhua (2006a) *Water Diversion Project Bill Sees 80% Rise*, available online at http://news.xinhuanet.com/english/2006–05/04/content_4508028.htm

Xinhua (2006b) *Yangtze River Becomes World's Busiest Freshwater Route*, available online at http://english.people.com.cn/200601/06/eng20060106_233300.html

Part Three – National Level

The interaction between basin levels and national frameworks is not dissimilar to that between local and basin levels. Many activities can be undertaken at the basin level, but there will be issues that can only be addressed at national level.

In the case of the Yangtze in China, the importance of the basin for various dimensions of national development (ranging from transport and clean energy to water supply to other basins and regional development and national security) meant that decisions about basin-level development necessarily involved the national-level.

In Denmark, the situation was somewhat different. While many individual communities were trying individually to address the diffuse pollution from agricultural activities and the local water resources were not interconnected, this local problem was generic across many areas of the country and was best dealt with in a national framework. This can be important where it is necessary to avoid the environmental equivalent of a 'race to the bottom' between localities competing to attract economic activity and fearful of the response if such activity is regulated or controlled.

All these issues must be taken into account in policy approaches at national level, as illustrated in the cases of both Chile and South Africa. The importance of the national framework and institutions rises in direct proportion to the degree of interconnectedness between communities and river systems. This is greater in South Africa than in Chile, with the result that water users have more autonomy in the latter.

But, as the Angas Bremer case demonstrated, as social and economic development places growing pressures on the resource, local approaches may need national responses, and it was significant, in this context, that water management was one of a handful of issues over which the historically fiercely independent federal states agreed should be overseen by their Commonwealth government.

11

Taking it One Step at a Time: Chile's Sequential, Adaptive Approach to Achieving the Three Es

Humberto Peña[1]

Chile's move towards more sustainable development and management of its water resources has been one of gradual adaptation, closely tied to the country's economic development. In the mid-1970s, Chile pursued a development model based on three major themes: maintaining macroeconomic equilibrium; strengthening the role the market plays in the allocation of resources, including water; and opening up the economy to world markets while exporting products for which the country had an advantage, almost all of which use water in their production processes.

This approach led to increased water use, but also increased economic efficiency and increased private investment in Chile's sewerage and water supply sector. Gradually, water policy was adapted to better address social equity and environmental sustainability concerns. This reflected a shift in governing ideology and an increasing public awareness of social and environmental issues.

Chile's example suggests the wisdom of developing water resource planning and management strategies in relation to the specific development model in place, and of using a phased approach to water-resource management – as opposed to policies that call for immediate integration of all water-related planning and management. As Chile's experience has demonstrated, economic development resulting in social benefits may enable environmental objectives to be met in the long term.

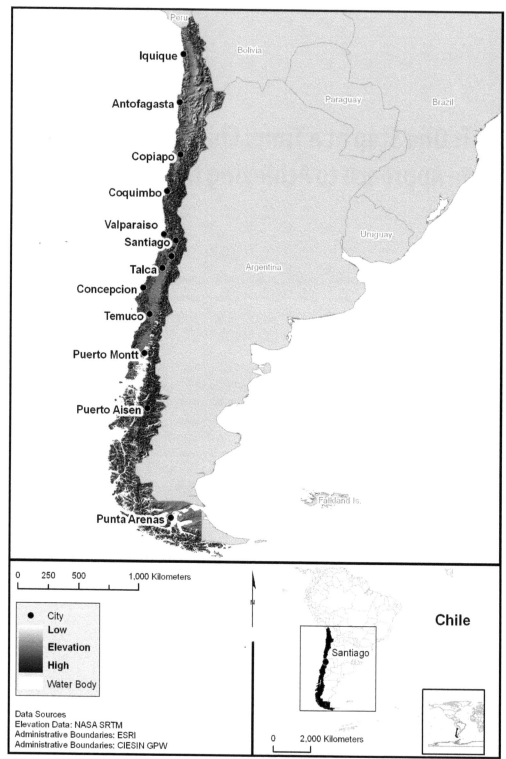

Figure 11.1   Map of Chile

The development context

Continental Chile stretches out along a narrow strip of territory, 4,200 kilometres long with an average width of 180 kilometres, located between the Andean mountain range and the Pacific Ocean in the southern cone of South America. In this territory, 200 small basins, which drain the west slopes of the Andes, are home to some 15 million people.

From the administrative standpoint Chile is a unitary republic, divided from north to south into 13 regions. In terms of water management, at the national level the General Water Directorate (DGA), the National Commission for Environment (CONAMA) and the Superintendence of Sanitary Services (SISS) are the main agencies. The DGA has offices in each of the 13 regions. In addition, there are water users' organizations (WUOs) operating at basin and sub-basin levels.

The availability of water resources in Chile is characterized by an extraordinary heterogeneity in spatial distribution. In the northern half of the country arid conditions prevail; the average water availability is below 1,000 cubic metres per inhabitant per year (in some regions as low as 500 cubic metres per inhabitant per year) – levels that international standards classify as very restrictive for development. To the south of Santiago, the capital city, which is located in the centre of the country, the availability exceeds 10,000 cubic metres per inhabitant per year (DGA, 1999).

Irrigated agriculture accounts for 85 per cent of the total consumptive water demand; approximately 1.3 million hectares of Chile's farmland is irrigated. Domestic uses account for 5 per cent, and mining and industrial uses share the remaining 10 per cent (DGA, 1999).

The level of competition between these uses varies throughout the country. In most of the country there is intense pressure on existing water resources. This pressure is particularly acute in the northern and central areas, where, since the mid-1900s, all surface water has already been allocated.

Therefore, making water available for new activities requires one of the following: an improvement in water-use efficiencies, especially in the case of irrigation; the exploitation of available groundwater resources; the building of reservoirs to store surplus water resources from the winter, spring or humid years; the reuse of treated water resources; or the desalination of brackish river water or sea water.

The growth in export agriculture posed a particular challenge for the water sector, since it did not develop in areas where there are plentiful water resources, but in the north and the centre of the country. The production of fruit and vegetables in these areas requires water for irrigation – water that was used in traditional crops. On the other hand, traditional agriculture – which focuses on growing crops for domestic consumption or import substitutes, such as wheat – is located primarily in the south and is primarily rainfed.

At the beginning of 1980s, the country had to face restrictions related to water quality and environmental protection and restoration issues. Chile had amassed significant environmental debts (commonly known as 'environmental liabilities') during the 20th century. For example, sewage was discharged without treatment into the rivers – causing high levels of gastrointestinal and enteric diseases downstream from the main cities and towns. In addition, there were no controls on the disposal of mining and industrial effluents.

At the end of the 1970s, radical economic reforms introduced by a new authoritarian government had a major impact on the water sector. These reforms were implemented based on the following major axes:

- An economic policy directed at maintaining macroeconomic equilibriums, principally by controlling the fiscal deficit and reducing inflation.
- A policy of opening up the economy to international trade, allowing the export of products in which the country is competitive and importing products in which the country is not.

This process, which was initially implemented unilaterally by Chile, has been complemented, more recently, by the signing of various trade agreements and free trade treaties with commercial blocks including the USA, the European Union, Japan, China, South Korea and Latin American countries, among others.

- The implementation of a series of market reforms in various sectors of the economy and a policy of reducing the activities undertaken by the state, in the sense that the state transfers those activities that can be undertaken by the private sector and redirects its efforts to regulation, the promotion of development, and to those activities that cannot be assumed by the private sector.

This strategy, reflecting the ideology of the new government, redirected the productive structure of the country towards dynamic development of exports based principally on natural resources.

It is very important to emphasize that nearly all of the products on which the Chilean export model is based involve the use of the water resource in their production (see Figure 11.2). For example, copper mining uses water in processing the metal once it has been mined. Moreover, water is scarce in the north of Chile and it is precisely there – in the Atacama Desert – that the major proportion of copper is produced. The production of fresh fruit is concentrated in areas where plants cannot grow properly without irrigation. Forestry plantations do not need irrigation, but water is used in the subsequent industrial processes of the lumber industry, such as in pulp production. Finally, salmon farming requires fresh water in the early stages of the growth of the fish.

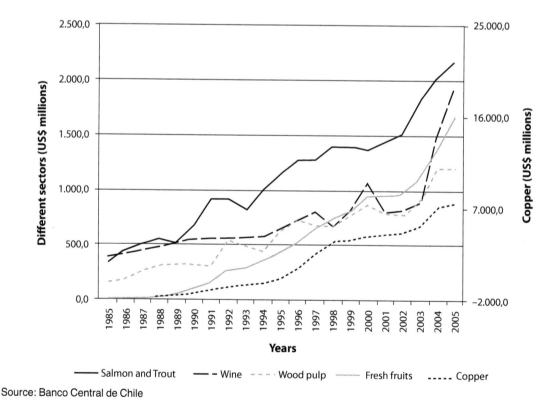

Source: Banco Central de Chile

Figure 11.2 Chilean exports based on water resources

Consequently, it can be said that, in general terms, good water resource management is fundamental to the success of the export model since those sectors of the economy dependent on water resources generate a high proportion of total exports. So, this model of economic and social development has affected the use and management of the water resource and has required changes in public policies related to water.

The approach

The Chilean case clearly demonstrates that water policies are the result of a complex interplay of influences within a society. Of these influences, some originate within the water sector itself, but most of the principal ones come from other areas of society, especially the predominant ideological perceptions and the choice of national development strategy. Obviously, the internal dynamics of the water sector can either favour or work against policy adjustments. On many occasions, it is these dynamics that determine whether certain changes are possible or not.

From this last perspective, in Chile from 1975 to 1990 there was a period in which the policies were designed under an authoritarian government, with a decidedly neo-liberal view and a total confidence in the functioning of the market. Thus there was little regulation, the role of the state and the need for planning was minimized, and social demands, including those of ethnic minorities, and environmental concerns were largely ignored. During this period, the government focused on policies related to creating and deregulating markets and to establishing the predominance of private initiative, cost recovery for the self-financing of public services, and other such issues.

With the switch in 1990 to a democratic government, the focus shifted – policies tended to assign a greater role to the state, regulation of markets, environmental and ethnic concerns, and attention to social demands. Additionally, the idea arose of seeking alliances between the public and the private sectors for the provision of goods and services. However, despite this shift in focus, the basic economic structure developed under the authoritarian government was retained.

These distinct periods were reflected in the water sector both in institutional and legal initiatives and in the evolution of public investment. During the authoritarian period, the Water Law (1981) was promulgated with a decidedly market orientation, the foundations were established for private participation in water supply and sewerage services (1989), and the Promotion of Irrigation Law (1986) offered partial financing through government subsidies, but by and large irrigation initiatives were left to the private sector.

During the democratic period, social, environmental and regulatory considerations were added to the water resource system. For example, the Law on Indigenous Peoples (1993) was enacted and the Promotion of Irrigation Law was redirected towards peasant agriculture (1994); an important reform was undertaken in environmental matters, through the promulgation and implementation of the Basic Environment Law (1994) and the development of a series of water pollution standards (1998–2002); and the reform of the Water Law (2005) was approved and a more balanced concept of water policy emerged, weighing the economic, environmental and social dimensions of water and reaffirming the regulatory role of the state (Peña, Luraschi and Valenzuela, 2004).

So, overall, the regulatory and institutional system designed for water resources at the beginning of the 1980s, conceived basically to give priority to the economic aspects of the resource and the application of market mechanisms to the sector, has evolved towards a more balanced view. This is a result of the challenges arising from the practical reality of water management and of the developing perception in different sectors of the role of water in national development, a change favoured by political and ideological change in the society as a whole.

At the end of this process, the functions of water management are shared between the state and the private sector (see Figure 11.3). The state has assumed the following functions related to water resources:

- Research and measurement of the water resource.
- Regulating water use to prevent overuse and damage to the rights of third parties and assuring environmental sustainability.
- Regulating services related to water resources and promoting conditions for their economically efficient development.
- Conserving and protecting the water resource within the context of environmental sustainability.
- Promoting the satisfaction of the basic needs of the poorest sectors of the population.

- Promoting, arranging for and, to the extent that there are social benefits, assisting in the financing of irrigation projects.

The private sector has the following responsibilities:

- Studying, financing, deciding on and undertaking development projects related to the water resource, including water supply and hydroelectricity projects. For these purposes water-use rights are treated as part of their commercial capital and water can be considered an input to the production process.
- It is the job of the private sector, organized into user organizations, to distribute the resource according to the rights of each individual and to maintain the common distribution infrastructure.

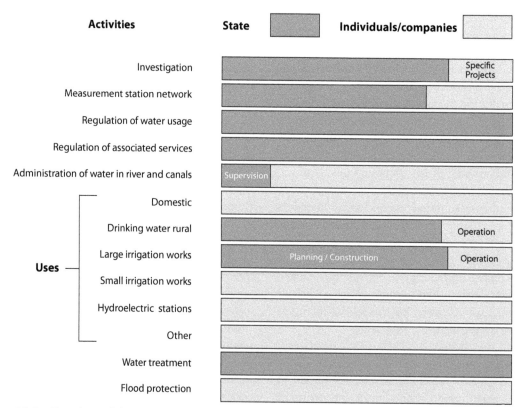

Figure 11.3 Functions of the state and the private sector in relation to water resources

Instruments used

According to this conceptual approach an institutional framework of water management has been built. For this purpose, the main instruments are the Water Law and the Basic Environment Law. Other important instruments are the systems of regulations and incentives governing other activities in which water is an input (hydroelectricity, domestic use and irrigation), such as the laws and regulations of the water supply and sewerage sector, and the law promoting irrigation.

The institutional system related to water resources includes public agencies with the following functions: regulation of withdrawals, regulation of discharges, environmental regulation of water use, regulation of public services, and assistance to irrigation. The main agencies are: the General Water Directorate (DGA), which is responsible for implementing the Water Law; the National Commission for Environment (CONAMA); and the Superintendence of Sanitary Services (SISS).

Water Law

In Chile, the institutional system for water is directed at achieving more economically efficient water use and allocation.

With respect to this objective, there is widespread national agreement regarding the benefits of using the market to reallocate existing water rights, and the need to establish property rights over water-use rights to provide legal certainty to water-related investments and to enable the market to reallocate water resources. On the other hand, the amendments recently approved by a great majority are intended to reconcile, in practice: water as a 'national property for public use' with the guarantees of property rights over water-use rights; economic incentives and competition with protection of the public interest; and the state's role in managing a complex resource so crucial to development with the promotion of private initiative and management transparency.

So the Water Law, with its amendments, provided for the market to play a crucial role in two areas: reallocation of water among private individuals; and original allocation of water rights. In terms of the first issue, the Water Law established that, although water would still be considered national property for public use, the rights to use water would have characteristics of property under civil law and would be the main object of rights, rights in themselves and not accessories to any other rights, and would be freely transferable.

With reference to the second aspect – original allocation of water rights – original water use rights would be allocated by the state and, in the event of two or more requests for the same water and insufficient availability to grant them all, water rights must be allocated through auctions. However, the President has the authority to protect the public interest by excluding water resources from economic competition when they need to be reserved for public supply in the absence of other means of obtaining water or, in the case of non-consumptive rights, in the event of national interest. Also, rules have been established to limit requests to genuine project needs to avoid speculative acquisitions. Similarly, the legislation states that the authority is obliged to consider environmental aspects in the process of establishing new water rights, especially in terms of determining ecological water flows and protecting sustainable aquifer management.

It is important to note that before the reform of the Water Code approved in 2005, ending 13 years of negotiations in the Chilean Congress, the state could not reserve water resources to protect public interest, users did not have to justify the quantity requested for new water rights allocation, right holders had no obligation to use the water, and environmental aspects were not considered in the process of establishing new water rights. This method of allocating water resources did not have the expected results, however, as the auction mechanism was hardly ever used in practice, and the allocation of water rights without any limits and restrictions gave rise to various situations that were detrimental to the country, such as the accumulation of water rights for hoarding and speculation, as barriers to entry for competitors in various markets and in order to preclude allocation of water rights to those who really needed them. One example was in the area of water rights for non-consumptive use (hydroelectricity), where 50,000 cubic metres per second were requested, an amount that is out of all proportion given that it could not possibly be put to use for several decades.

So, with the reform, without prejudice to environmental considerations and the reservation of water resources in accordance with the public interest, the allocation criterion for choosing between various requests will tend to be mainly economic, in practice, given that it is in the country's interest to allocate scarce water resources to those activities with the highest productivity per cubic metre of water. Unlike in other countries, there is a general consensus between water users and political sectors in Chile that it would be unwise to give preference to the requirements of a particular user sector, on the basis that this would encourage inefficiency and fail to signal to users the relative scarcity of the resource.

Basic Environment Law

The Basic Environment Law was promulgated in 1994 and is the principal body of law currently regulating the environmental management of the water resource in Chile. It established a number of related means for protecting the environment and preventing and controlling pollution, the most important amongst them being:

- Emission and environmental quality standards have been issued for both surface water and groundwater relating to the health of the population and the state of ecosystems.
- Plans have been put in place for pollution prevention, if the pollution is about to surpass the standards, or for pollution clean-up measures, if the standards have been exceeded.

- The environmental impact evaluation system (SEIA). This system evaluates large investment projects, both public and private. It applies to projects that affect the environment, according to criteria established in the law, and gives environmental authorization, while suggesting improvements to the environmental profile of the projects (mainly mitigation measures, compensation, restoration and prevention). A committee constituted by the agencies with environmental functions evaluates every project. Also, public participation is considered in the process.

It is important to note that environmental policy for water resources has tended to evolve jointly with water supply and sewerage policy, due to the need to improve pollution conditions and reduce health problems. Consequently, the first environmental standards issued have been in reference to water supply and sewerage, establishing gradual goals for sewage treatment.

Water supply and sewerage sector

Up to 1989 the responsibility of supplying water to the population remained fundamentally in a single national service under the Ministry of Public Works. Starting from 1989, a deep transformation of the sector was carried out, creating, based on the existing public service, 13 regional companies, one for each region of the country. Additionally, a Superintendence of Sanitary Services (SISS) was organized, which assumed the regulating tasks of the state, especially relative to the quality of the given services and the rates charged for the service.

A characteristic of the stage of the state as regulator is the separation of the normative and monitoring functions from the functions of production and commercialization of the services. So the Chilean policies governing the water supply and sewerage sector are that the production, distribution, collection and treatment of water are undertaken by those companies, both public and private, that are regulated by the state through the tariff system and the control of their investment plans. According to the legislation, the value of the tariffs must reflect the real cost of providing the service including an adequate return for the company. For its part, the state assists the access of the less advantaged population through targeted programmes. In urban areas the state subsidizes supply to the poorer population through a municipally run system of partly paying their bills, while in rural areas water supply is subsidized through a specific programme under the responsibility of the Ministry of Public Works.

Finally, in 1998, the Law was modified to strongly reinforce the regulating role of the SISS as, in practice, especially in drought periods, the instruments available to enforce the responsibilities of the companies to provide adequate service had proved to be insufficient. This legal modification was also conceived of as an indispensable step prior to the massive incorporation of the private sector into the management of the companies.

Since then, in accordance with the environmental law, efforts have been directed towards controlling water pollution with an integrated and long-term view, as well as towards solving the major problems rapidly through the construction of waste treatment plants.

Irrigation sector

In Chile, the state promotes and coordinates projects for medium and large works that the private sector would find difficult to undertake because of the tremendous coordination required for the study, expropriation and building of the works. With regard to finance, the ruling policy is that the beneficiaries pay a part of the capital invested once the works have been built, although in practice this has not led to a high rate of cost recovery.

To promote minor on-farm works, a policy of direct subsidy to the private sector has been developed. The drive to enhance on-farm irrigation and to stimulate the building of small water works for agricultural use began in 1986 with the Promotion of Irrigation and Drainage Law. This law, which includes projects such as modernization of irrigation systems, projects for the improvement of transport and small reservoirs, offers subsidies for projects that cost less than approximately US$350,000. The system operates through public competitive bids and in practice the amount of the subsidy is about 50 per cent, depending on the number of projects in each competition. Because it was not possible for poorer farmers to compete with commercial agriculture for the funds assigned under the original law, in 1994, the law was amended in order to have separate competitions for 'commercial agriculture' and for 'poorer farmers'.

Integrated water resources management (IWRM) and water policies

The institutional changes that have occurred since the late 1970s in Chile must be understood as a process to build, step by step, an institutional framework inspired by IWRM.

The main driver in this process has been the need to balance economic growth, based on products highly dependent on water resources in a context of scarcity, with environmental and social goals.

For this purpose, the institutional framework takes an integrated approach to:

- national goals and water policies;
- water availability and demand;
- water allocation between different sectors and users; and
- environment and water management.

National goals and water policies

In practice, when water problems seem to be relevant for the success of public goals, the convergence of national goals and water policies has been established at the highest level of government

– for example, in the case of policies regarding promotion of exports and investment, public health, agriculture development, social equity and others. To solve these problems a wide range of instruments has been used. So, in some cases, the main role has been assigned to market incentives and the private sector by law (export policies); in others a regulatory approach has been used, defining standards to be achieved (pollution control, health policies), and policies on subsidies have been implemented in some areas (water supply for rural areas, domestic uses for poor people, irrigation); and in others, direct intervention by the state has been implemented (large irrigation projects). It is important to note that there is not a permanent system to define water policies. Obviously, this approach has several drawbacks; however, in practice, the political system has been sensitive enough to the role of water management. Therefore, institutional changes have been promoted by certain ministries (including the ministry of finance), by state policies lasting several governments or by coordination committees between different ministries and agencies, among other alternatives.

Water availability and demand

The increase in productive activities related to exports has created a situation in which it is necessary to coordinate decisions about demand for water with water availability. In the Chilean case, this challenge has been tackled by creating an institutional system where a water market gives signals about water scarcity through the price of water rights. Also, due to the judicial security of water rights, greater scarcity and increases in the water price provide incentives for the private sector to invest in improving water-use efficiency (see Figure 11.4). So, around the country, according to each company's analysis of its local reality, different supply strategies are defined. There are some exceptions to this general approach, when the state has subsidized the demand for social reasons (water supply for poor people).

Water allocation between different sectors and users

The state determines the total water rights in a river or aquifer compatible with environmental sustainability. When there is no longer water available to meet all the demands, water rights are allocated between the different sectors and users through markets, with only few exceptions due to social reasons. Also, because the water rights are independent of a specific use, they can be transferred between sectors in the water market. As it was pointed out, the physical availability of water resources, as well as the specific demand for different

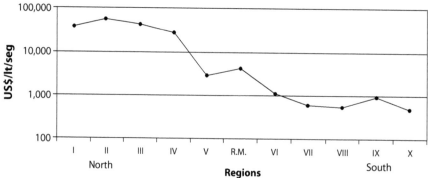

Source: Author, based on SISS data

Figure 11.4   Water rights prices

sectors, is very variable around the country, and consequently the interest in the water market will depend fundamentally on the cost of available options and on other operational considerations relative to safety of supply. So when water is scarce the allocation system is designed to allocate the water to the most productive use.

However, before the approval of the water code reform, several distortions were observed as a consequence of the original water rights allocation system. Also, there are some institutional failures that need to be addressed to make market reallocations more effective (see 'pending challenges' below).

Environment and water management

In Chile, initiatives dealing with water and environment are integrated, mainly, in the framework of the National Commission for Environment (CONAMA), which is a coordination entity where different sectors and agencies are represented, including those related to water. CONAMA's highest authority is a council of ministers, so some of the policies are analysed and implemented directly, in a participative way, by CONAMA. For example, in the cases of the environmental evaluation systems and the process of defining water quality standards in rivers. In others cases, water regulations and policies, such as those related to Water Law and sanitary services, are implemented by different agencies (DGA, SISS) with specific environmental functions.

According to the IWRM approach, these institutional solutions should be aimed at ensuring economic efficiency in water use, with attention to social equity and environmental sustainability concerns.

In the Chilean case, these objectives are addressed in the following way:

- *Economic efficiency in water use*: In the Chilean model, the main drivers of improved efficiency are market incentives. In fact, through the price of water rights, users have a signal to invest in avoiding water losses, and to allocate the water rights to highest value uses. Also, the use of new technologies, with higher efficiencies, is promoted because they are fundamental to producing quality goods that compete successfully in world markets.
- *Social equity*: Water policies have several social goals that are incorporated into laws or governmental programmes, in order to make access to water more equitable for everybody. So, in some cases the bidding system for water rights allocation is not used, water rights used by native people are recognized by law, and, in order to reduce the impact of expenditures on basic services, a policy of focalized subsidies for the lowest income groups has been adopted. Also, focalized subsidies have been developed for the consumption of electricity and irrigation for small farmers.
- *Environmental sustainability*: As it has been mentioned, the Environmental Law, Water Law and water sewage regulations include several instruments for protecting the environment and controlling its pollution. Among these, the most important are the rules established in order to preserve ecological flows in new areas, protect wetlands, define standards for water discharges and water quality in rivers, and evaluate the environmental impact of new projects.

Pending challenges

Even if the basic institutional structure can answer the main social, economic and environmental demands, the following challenges are still unresolved.

Integrated basin water resources management

While the model adopted by Chile has been successful in promoting private investment in natural resources, it has not been as successful in encouraging dialogue and negotiation between the different water users in a basin, especially when pressure on water resources has become higher. Most of the water users' organizations were created in terms of a law passed at the beginning of the 20th century, by the landowners. Since the organizations were first formed, the users have changed, but the WUOs have maintained the same structure and continue as autonomous entities without government intervention. As the WUOs represent only those with formal water rights, they do not necessarily represent all the stakeholders, and have limited ability to engage groups who impact water resources in a basin (e.g. polluting industries, cities) and those who are impacted by water-use decisions (e.g. fishers, recreational water users).

There are no effective mechanisms at the basin level to negotiate and solve issues affecting the whole basin, such as erosion, floods and diffuse contamination. Also, several conflicts between consumptive water uses (e.g. irrigation) and non-consumptive water uses (e.g. hydroelectricity) have been generated in the past. There is poor utilization of the advantages of a conjunctive management of surface water and groundwater. Lack of consensus about future scenarios creates difficulties in relation to public and private decisions. Finally, the current system for evaluating environmental impacts does not adequately consider the cumulative impact of different projects, at the basin level.

Environmental institutions

A second problem regards the implementation of environmental law and its regulations. In 1994, public agencies, the private sector and other stakeholders had neither experience nor specialist capacity in environmental impact evaluation, and environmental information and knowledge, especially on aquatic ecosystems, were poor. Besides, institutional design did not distinguish clearly between political and technical roles. So the environmental review prepared in the OECD

framework (OECD, 2005) recommended developing and strengthening environmental institutions at national level and converting them to an autonomous public regulatory and coordination body, and reviewing ways to strengthen compliance and enforcement capacity, including through institutional reforms such as the establishment of an environmental inspectorate. The government is considering these recommendations and Congress is discussing a law to create an Environment Ministry and a Superintendent of the Environment.

Strengthen water institutions' capacity

Institutions, both public agencies and water users' organizations, need greater capacity to enforce the water code regulations, especially for aquifers, and provide the databases, procedures and knowledge to support good water-market performance. There is no doubt that this is a long-term task, because several capacities at different levels must be developed.

Outcomes and impacts

Support to export policies

Export policies offered a major challenge for the water sector: how to allow the increases in exports of several products with high water demands, despite the fact that available water resources were committed. That the sector has met this challenge is demonstrated by the statistics: water-intensive exports have increased more than eightfold since 1985 (see Figure 11.2) (Banco Central de Chile, 2008). For example, fruit exports have grown sevenfold, exports of wine have increased more than fiftyfold, mining has expanded ninefold, and forestry products, mainly pulp and sawn lumber, have increased eightfold. If the country's water policies had not changed, Chile would not have been able to obtain these results (ODEPA, 2005).

Water-use efficiency and productivity

The general change in social and economic activity has not only influenced the total volume of water required, but also the way in which it is managed – enforcing the improvement of efficiency and productivity of water use. This is a consequence derived from greater scarcity, higher prices of water rights in water markets and from other economic forces associated with the country's foreign trade.

The increase in irrigation efficiency has been achieved through the incorporation of new technologies. In many cases, the new technologies were adopted not to reduce water consumption, but simply as a technological standard for investment projects, particularly those associated with exports, considering that such technologies are fundamental to the development of quality products that compete successfully in world markets. Adoption of these technologies has been favoured by state subsidies through the Promotion of Irrigation and Drainage Law. So, at present, almost 30 per cent of irrigated land uses highly water-efficient technologies (INE, 2008).

At the beginning of the 1980s, mining developed in a context of little competition for water and, consequently, the efficiency of use was low – an estimated three cubic metres of water for each tonne of mineral processed. More recently, scarcity and the price of water rights provided an incentive to increase efficiency in water use. So, by 2000, average consumption reached 0.75 cubic metre for each tonne of mineral processed. (Figure 11.5 shows the case of the second largest mine in Chile.) Also, significant increases in efficiency have been achieved in wood pulp production: today, only 40 cubic metres of water are consumed for each tonne produced, while in the 1980s consumption was around 130 cubic metres per tonne (Peña et al, 2004).

When water is used more efficiently in production processes, the resulting productivity of the water resource will be higher. The economic productivity of water can also be increased when water is reallocated to production of higher-value goods, when new demands are satisfied by transfers of water rights only marginally used, or when the production structure of agriculture switches to

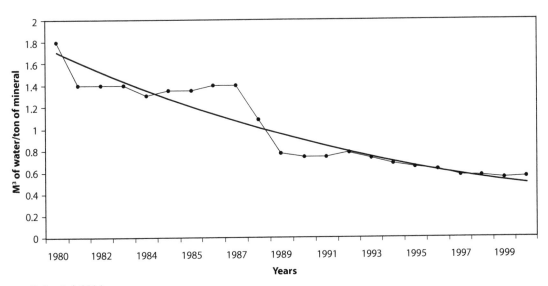

Source: Peña et al, 2004

Figure 11.5   Water use efficiency – Chuquicamata Mine

higher-value crops which use less water. This is the case, for example, in the valleys in northern Chile, where the economic productivity of water increased by 58 per cent over a period of ten years (CAZALAC/RODHOS, 2006).

Water supply

At present, in Chile the indicators of water supply and sanitation services are at the level of those of developed countries. Potable water supply coverage has reached 99 per cent of the 14 million inhabitants that live in urban areas, and 85 per cent of the 1.5 million rural inhabitants. In urban areas, sanitation has reached 85 per cent of the population (SISS, 2008).

Pollution and the environment

In relation to pollution control, the main achievement of water policies has been the investment programme in sewage treatment, which has developed since 1998, in accordance with environmental law and regulations. Pollution has decreased substantially since 1998, when only 17 per cent of the sewage was treated, to 80 per cent in 2007. By 2010 this source of pollution is expected to be almost 100 per cent controlled (see Figure 11.6) (SISS, 2008).

Also, the approval of standards for the discharge of liquid industrial wastes have had a positive

influence on the decision to build treatment plants, increasing the number of plants fivefold since 1998. However, as mentioned above, water quality control is still weak.

Significant improvements are not observable in the areas of pollution from nutrients and pesticides by diffuse sources and environmental liabilities originating in old mining activities.

Chile has the lowest rates of infant mortality and mortality among children under five in Latin America, with rates comparable to those of developed countries. There can be no doubt that the water supply and sewerage services have made a significant contribution to the health indices achieved (CEPIS, 2000).

In relation to the environmental impact of investment projects, every year, since the Basic Environmental Law was approved, more than 1,000 projects – 50 per cent of which are related to water – are evaluated, with a projected investment of around US$5 billion (CONAMA, 2008). Despite this, concerns have been raised about the quality of the evaluations, which is one reason why further reforms of environmental institutions are being considered.

Lessons learned

The Chilean case shows the interrelationship between the growth of the country, its demand for water, and the emergence of environmental concerns related to the use of natural resources. Consequently, it demonstrates the importance of integrating water resources policies and those for economic and social development, for reaching national goals.

The evolution of the institutional system is the result of complex societal influences and the challenges that have arisen along with economic growth. It is not the result of a unique design or of a structure that was coherently conceived from day one; rather, it is the result of a process of improving the original design and addresses weaknesses. The

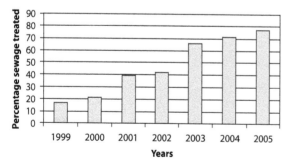

Source: SISS, 2008

Figure 11.6   Evolution of sewage treatment

above suggests that, in specific situations, a step-by-step strategy for change in the water sector can be quite enough – particularly if it is governed by a realistic and pragmatic strategy of resolving the urgent and the possible, rather than calling for global reforms.

Policy implementation requires resolution of the financing question. In poor societies this issue is critical, since the great majority of beneficiaries of policies are not able to effectively contribute to their financing, and the general resources of the state are usually insufficient due to the enormous pending demands of other kinds.

In this case, Chile shows a very clear example of a rigorous phasing of social objectives in the water sector, even though this was probably less by design and more a reflection of the general political change in the country.

This analysis suggests that proposal for new policies and social goals should be kept compatible with the real capabilities of the country, particularly in the design of methods by which they will be financed.

There is a common tendency to see a contradiction between environmental conservation and the economic development process. The Chilean experience of this question in the water sector, although there is no evidence on which to base definitive conclusions, does not appear to validate such an affirmation. It is true that threats to the environment increase as a result of the greater demand for natural resources; for example, for use of groundwater and for the use and/or production of potentially polluting materials. However, there is also an increase in investment in pollution control and a greater attention to the design and implementation of policies directed towards the control of environmental impacts. As has been mentioned, in this interplay the globalization process itself plays an important role in that it allows the transfer of experiences and technologies from more developed countries, and works towards the establishment of international standards. Moreover, as has been mentioned, in many cases competition

in global markets presupposes a high degree of technical skill in water management and the necessity of international certification to assure access to other markets.

Finally, it is important to note that Chile has never had an explicit 'IWRM policy' nor did the country ever set out to 'implement IWRM'. The process described here was one that set out to solve water problems as they arose and to ensure that water did not become an obstacle to social or economic development. To do this required an integrated approach and one that eventually encompassed social equity, economic efficient and environmental sustainability – the principles of IWRM. This is an ongoing process and in some areas there is a need for more rapid progress. But even as these pending challenges are met, new ones will arise. Such is the continuous process of IWRM.

Note

1. Humberto Peña served as Director of Chile's Directorate of Water Resources from 1994 to 2006. As such, he played a principal role in the definition of the National Water Policy and the Chilean Water Reform.

References

Banco Central de Chile (2008) *Statistics*, available online at www.bcentral.cl

CAZALAC/RODHOS (Water Centre for Arid and Semi-Arid Zones of Latin America and the Caribbean) (2006) *Aplicación de Metodologías para Determinar la Eficiencia de Uso del Agua*, Direccion General de Aguas, Coquimbo

CEPIS (Pan American Centre for Sanitary Engineering and Environmental Sciences) (2000) *Assessment of Drinking Water and Sanitation in the Americas*, Chile Analytical Report, CEPIS, Lima

CONAMA (National Commission for the Environment) (2008) *Statistics*, available online at www.conama.cl

DGA (General Water Directorate) (1999) *Política Nacional de Recursos Hídricos*, DGA, Santiago de Chile

INE (2008) *Censo Agropecuario 2007*, Instituto Nacional de Estadisticas, Santiago de Chile

ODEPA (Office for the Study of Agricultural Policies) (2005) *Agricultura Chilena 2014. Una Perspectiva de Mediano Plazo*, ODEPA, Santiago de Chile

OECD (Organisation for Economic Co-Operation and Development) (2005) *OECD Environmental Performance Reviews – Chile*, OECD, Paris

Peña, H., Luraschi, M. and Valenzuela, S. (2004) *Water, Development and Public Policies*, South American Technical Advisory Committee (SAMTAC), Economic Commission for Latin America and the Caribbean (ECLAC) and Global Water Partnership (GWP), Santiago de Chile

SISS (Superintendence of Sanitary Services) (2008) *Statistics*, available online at www.siss.cl

12

Attempting to Do it All: How a New South Africa has Harnessed Water to Address its Development Challenges

Mike Muller[1]

South Africa's transition in 1994 from oppressive minority rule to a vibrant non-racial democracy was hailed as one of the political successes of the 20th century. With the political transition underway, attention turned to the many challenges of economic and social development faced by the country and the environmental constraints within which they had to be met.

Water was one of these constraints. With national water use approaching total water availability – and exceeding it in some catchments, to the detriment of the environment – it was recognized that there was an urgent need to move from developing the resource to focusing on its effective management. In this context, South Africa has had to attempt to 'do it all', to address social, environmental and economic needs in a single, sweeping reform process.

The key lesson from the South African experience is that broader political change can create important opportunities to improve water management. The reform is more likely to be successful if it is sequenced to address social objectives at an early stage to maintain political support for more complex economic and environmental reforms.

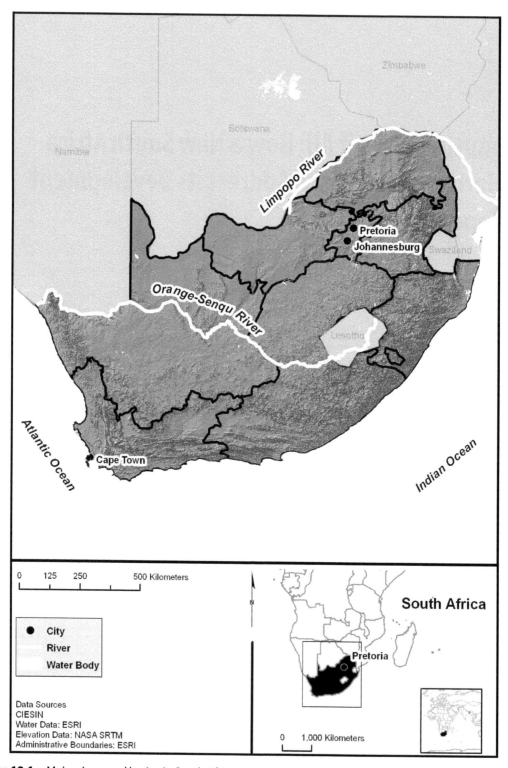

Figure 12.1   Major rivers and basins in South Africa

The development context and South Africa's water challenge

If the volume of renewable fresh water available per person annually is used as a measure, South Africa is one of the 30 driest countries on earth (UNESCO-WWAP, Table 4.3, 2006). Available water is unevenly distributed, with half the country largely semi-desert, and the climate is extremely variable. In addition, the country's main rivers, notably the Orange-Senqu and the Limpopo, are shared with six neighbours whose resources are also limited.

Water has been crucial to South Africa's economy from the time when Holland colonized the Cape as a watering and victualling station for the Asia-bound fleet of the Dutch East India Company. In many areas, the expansion of agriculture depended heavily on appropriating the limited natural water resources. And when, in the 19th century, the gold and

diamonds that formed the basis of South Africa's industrial economy were found, new pressures were placed on the water resources.

Since the economy developed around the exploitation of the country's mineral resources, a large proportion of the population and 60 per cent of its national GDP are concentrated inland, at the centre of the country and near the source of a number of important rivers. Indeed, as the Government's National Spatial Development Perspective has highlighted, 'the dominant pattern of settlement and economic activity in South Africa is largely out of line with water availability' (Presidency of South Africa, 2003: 13), as Figure 12.2 shows.

This poses challenges both for supply – water often has to be pumped uphill to the users – and for water quality – since pollution is generated at the top of the catchments, it impacts everyone who lives downstream.

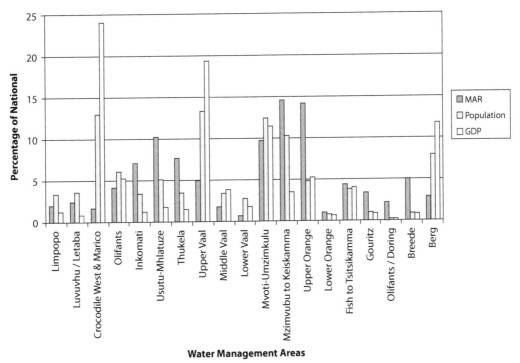

Source: DWAF, 2004

Figure 12.2 The regions with the highest contribution to GDP and the highest proportion of the population are often those with the lowest proportion of the water, as measured by the Mean Annual Runoff (MAR)

Reliable water supplies are already intensively used. One indicator of this is the extent to which the different river basins have been interconnected, to maximize their yields and to provide security against drought (Figure 12.3). Effective water management has long been recognized to be vital for national wellbeing and progress. And over the course of the 20th century, national competences were developed and institutions built to address the technical and investment challenges.

By 1990, despite the extensive infrastructure development which included enough dams to store 65 per cent of annual flows, water use was approaching the amount of water that could be made available economically and reliably at national level, and exceeding it in 10 of the country's 19 water management areas, to the detriment of the aquatic environment as well as to the users (see Table 12.1). There was thus an obvious need to move from the historic approach of developing the resource to increase supplies to focusing on effective resource management and conservation.

However, as in many countries, agriculture had long been the largest water user. And it was entrenched interests, notably the white farming community, that had blocked reforms, many of which had been proposed by the government's own water managers, backed by a national Commission of Enquiry, almost 25 years before (Government of South Africa, 1970).

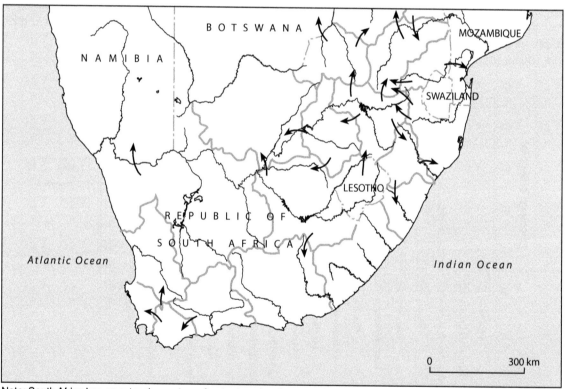

Note: South Africa has an extensive system of water transfers between rivers. Thus water that was flowing in the Orange River to the Atlantic may be transferred to the Vaal for use in Johannesburg and then be discharged as treated wastewater into the Limpopo River Basin, and thus flow through Mozambique and into the Indian Ocean instead.

Source: DWAF, 2004

Figure 12.3   Water management areas showing inter-basin water transfers

Unsustainable water use was only one of many challenges facing the new South Africa. The country also inherited a skewed and highly unequal economy from the minority-based government that was in place until 1994. The transition to democracy added to the challenges of the water sector. Newly established local government institutions had to immediately address water needs since one of the highest priorities of the newly enfranchised voters was safe water, conveniently available – something that had been denied to them in the past (Muller, 2007). The expansion of sanitation services put further pressure on the water resources.

However, political change also provided the opportunity for fundamental reforms in the way water is managed. The transfer of political power weakened the entrenched interests that had previously obstructed change and unblocked the streams of discussion that were needed to pilot a more sustainable future (Muller, 1998).

This chapter reviews the policy process that was followed in South Africa, from the perspective of a policy-maker, and outlines the main reforms that were introduced, the policies and institutions that were developed, and the instruments and approaches used.

The focus is on the big ideas, the strategic interventions and the tactical innovations that made it possible to make rapid progress.

Finally, the quality of that progress is reviewed to answer the key question – did the structured approach that was adopted to improve water resource management really make a difference to the lives of ordinary South Africans?

The reform process

South Africa's water reform started with one great advantage: it was part of a much larger process. The new government came to power with an overwhelming political mandate for change. One small part of that was to 'undertake a process to involve all relevant parties in updating the Water Act to ensure the right of all South Africans to water security' (ANC, 1994: 30). There was no need to persuade decision-makers and stakeholders of the need to interrogate and reframe the way water was managed in the country because similar processes were happening in just about every corner of the country's life. This meant that there was a framework for policy review and institutional transformation that the water sector could adopt without having to bang too loudly on the table for attention. And, in the excitement of the transition, there was a great deal of energy and interest among ordinary people, as well as the major interest groups, in participating in such processes. All concerned went about this with gusto through a range of consultative processes that drew a wide community into a five-year period of debate, and produced a national water policy (DWAF, 1997) which guided the drafting of new water resource legislation (Government of South Africa, 1998).

While the structure of government had to be transformed to address the new realities and the needs of the majority of the population, an early decision had been made to retain the country's Department of Water Affairs and Forestry (DWAF) as a national organization to manage the water resources of the country. In addition, it had already been decided to expand DWAF's mandate to include leading the provision of water supply and sanitation services. This maintained not just technical competences, but also the institutional memory of a complex sector that has to prepare for extreme floods and droughts that might only occur once every 20 years.

There were some predictable problems. Counter-intuitively, one was that water was such a high priority in the many communities that had been denied access to safe water in the past (an estimated 12 out of 36 million people were without safe water in 1994; 21 million were without safe sanitation). From a water resource perspective, this meant that when water was put on the agenda of a meeting, most people wanted to talk about how to get safe water piped to their homes or those of their constituents; there was less interest in environmental protection, long-term water security and how water should be shared with neighbouring countries. On the other hand, since water resources had been monopolized by a minority of white farmers in the past, there was not a great deal of popular interest or sympathy in protecting the privileged water use enjoyed in the farming areas (Muller, 2007).

Key concepts

In the policy area, some of the 'big ideas' had been identified, if not implemented, in other countries.

First, the legal status of water was clarified; it was defined as an 'indivisible national resource', with national government responsible for its custodianship as a public trust. This legislation gave the national government a duty to regulate water use for the benefit of all South Africans, taking into account the public nature of water resources and the need to make sure that there is fair access to these resources, and that they are beneficially used in the public interest.

Then the objectives of water resource management, which often cause confusion, were clarified. In South Africa's Constitution, the distinction was drawn between the management of water resources (a natural resource in rivers and underground), which is a national competence, and the provision of water services (crudely put, managing water in pipes), which is a local government matter (Government of South Africa, 1996).

The Constitution, enacted in 1996, provided the legal framework within which property rights and environmental protection could be addressed in the new political dispensation; it also provided a right of 'access to sufficient food and water', although this has little impact on the management of water as a resource.

Other 'big ideas' that shaped the evolution of water management included:

- a clear statement of the goals of water resource management;
- establishing a formal basis for protection of aquatic ecosystems;
- treating underground water, as an integral part of the hydrological cycle, in the same way as surface water;
- transforming the concept of water rights from what were perceived to be permanent property rights to temporary use rights;
- recognizing that many elements of water management could more effectively be undertaken on a catchment basis;
- ensuring that water resource allocation would be guided by social goals; and
- moving beyond infrastructure development to focus on water conservation and effective management as well.

The first draft of the Water Law Principles had suggested that 'the core objective of water resource management ... is to ensure that water is available in sufficient quantity, quality and reliability for the development and well being of the nation'. That certainly had been the lodestone of South Africa's water managers in the past, engineers for whom building dams to make new supplies available was a rite of passage.

After some consideration, it was agreed that this statement did not reflect the new imperatives. Despite the best efforts of the engineers, the reality was that there was simply not enough water available to support water consumption if it continued to grow as it had in the past.

A more modest but arguably more radical vision was adopted, namely that 'The objective of managing the quantity, quality and reliability of the nation's water resources is to achieve optimum long term social and economic benefit for society from their use' (DWAF, 1996).

The new approach was less heroic, more humble, but crucially better reflected South Africa's future reality in which it was going to have to manage the resource and the tension between growing demand

and finite supply, and ensure that the limited water that is available is used not just productively and beneficially, but optimally.

The environmental reserve

The requirement that water be reserved to meet environmental needs represented a major advance, which water administrators and others had been seeking for decades. The approach taken was guided by the provision in the Bill of Rights that stated:

> Everyone has the right ... to have the environment protected for the benefit of present and future generations through reasonable legislative and other measures that: prevent pollution and environmental degradation, promote conservation and secure ecologically sustainable development and use of natural resources while promoting justifiable economic and social development.

This supported the introduction of the concept of the environmental reserve, which requires that, before water from a river or other source is allocated for other uses, the environmental requirements of that resource itself should be determined and 'reserved'. (A similar provision for a reserve for basic human needs was included; it was both more obvious and, in quantitative terms at least, of lesser impact.) An early estimate (see Table 12.1) was that approximately 20 per cent of flows would have to be reserved for environmental purposes (compared to less than 2 per cent for basic human needs).

Table 12.1   *Natural mean annual runoff and the environmental reserve (in millions of cubic metres per year), and storage in major dams (in millions of cubic metres)*

Water management area		Natural mean annual runoff (1)	Ecological reserve(1, 2)	Storage in major dams (3)
1	Limpopo	986	156	319
2	Luvuvhu/Letaba	1,185	224	531
3	Crocodile West and Marico	855	164	854
4	Olifants	2,040	460	1,078
5	Inkomati (4)	3,539	1,008	768
6	Usutu to Mhlatuze (5)	4,780	1,192	3,692
7	Thukela	3,799	859	1,125
8	Upper Vaal	2,423	299	5,725
9	Middle Vaal	888	109	467
10	Lower Vaal	181	49	1,375
11	Mvoti to Umzimkulu	4,798	1,160	827
12	Mzimvubu to Keiskamma	7,241	1,122	1,115
13	Upper Orange (6)	6,981	1,349	11,711
14	Lower Orange (7)	502	69	298
15	Fish to Tsitsikamma	2,154	243	739
16	Gouritz	1,679	325	301
17	Olifants/Doring	1,108	156	132
18	Breede	2,472	384	1,060
19	Berg	1,429	217	295
	Total for South Africa	**49,040**	**9,545**	**32,412**

Source: DWAF, 2004

Licences to use water, with limited trading

Another critical element of the new policy was to establish that water rights were use rights (rather than property rights); as such, they became subject to regulation in much the same way that land use is routinely subject to zoning. This recognizes that priorities for water use will change with time and provides a mechanism to reflect such changes. By stating that water use rights have a limited life span, the concept of fixed, permanent allocations was definitively ended.

At the same time, this was done in a manner that recognized the need for certainty and continuity amongst water users. This was vitally important. There was initial concern that the approach would undermine economic activity based on water. But negotiations with South Africa's major banks, as well as agricultural organizations, resulted in a helpful compromise: use rights could extend for a period of up to 40 years and the length of a licence would be determined, in part, by the extent and nature of the investments made in relation to the water use in question. No factory or farm could arbitrarily lose the water on which it depended for production.

Trading in water use rights was a contentious issue; there were real fears that some inland farmers would sell their water to mines and cities and retire to the coast, thus leaving government to solve the problem of tens of thousands of people stranded in suddenly unproductive rural areas. However, there was no desire to establish rigid administrative systems that would prevent local farmers from trading allocations between themselves. As a result, while all trades had to be registered, far stricter regulatory control was imposed over transfers between different user sectors and those that took water outside the individual sub-catchments.

Consultative allocation to achieve management goals

Internationally, experience has shown that, because water is such a high-volume, low-value commodity, it is difficult to establish effective markets to allocate

it. To address this, two forms of allocation were introduced. For individual applications for water use rights, clear criteria were established to guide decision-making. In catchments where there was already significant competition between users, a periodic allocation process, which includes compulsory licensing, was introduced. This process is to be managed by catchment stakeholders to reflect local priorities within the framework of the original licences.

While farmers and industrialists could not lose their water during the term of their use right, they might be challenged during a reallocation process to invest in greater efficiency if, through such investments, water could be made available to other users or to meet the needs of the environment. Thus large water users were given the incentive to improve their efficiency, since doing so contributes to their own water security, embodied in the extension of their licences.

Water conservation demand management

The notion that water conservation and demand management were projects with as much validity as dam construction and inter-basin transfers was one specific outcome of the policy debates of the pre-1994 period.

Water services programme

In parallel with these ideas and usually in the foreground, in the public perception at least, priority was given to an intensive national programme to provide safe water to the 12 million South Africans who did not yet have access to it (DWAF, 1994). The national programme reached more than 10 million people in the first ten years; housing and municipal programmes extended safe water to another 4 million (the number of people the government needed to provide service to increased over the period, along witht he population). The progress in this area gave credibility to and mobilized political support for water resources management efforts, particularly since the department responsible for water resources management, DWAF, was called upon to help new municipalities to establish their water supply and sanitation functions. Demonstrating that water reform meant practical improvements in the lives of poor people was a vital driver in the process. Innovations such as providing free basic water were part of a broader pricing policy for water services, which included the introduction of stepped tariffs that encouraged conservation by high volume users (Muller, 2008).

From these big ideas, policies and strategies, flowed a range of instruments and institutional arrangements to put them into practice.

From policy to implementation with equity, efficiency and sustainability

Good ideas remain just that, ideas, until they are implemented. And implementation is often the hurdle at which many good ideas, including sound water policies, fail. So a clear implementation strategy is a precondition for any successful policy.

The 1998 National Water Act requires the preparation of a National Water Resource Strategy, which is to be reviewed on a regular basis, to guide the activities of all parties responsible for water resource management.

The Strategy has to be developed in consultation with all interested parties. Once prepared, it establishes a statutory framework within which local and regional management organizations have to work – determining, for instance, how much water has to be released to neighbouring countries or transferred to other catchments. As important, it enables water users from all sectors to understand how they fit into the bigger water picture.

The first edition of the Strategy, promoted initially as South Africa's 'blueprint for survival', was approved and published in 2004. It set out a 20-year programme of implementation that includes the establishment of new regional water management organizations, the new infrastructure investment programme and, critically, the phasing in of the new management approaches. It incidentally made South Africa one of the first countries to meet the target for the development of national IWRM and water efficiency strategies set at the 2002 World Summit on Sustainable Development.

In both the National Water Act and the Strategy, equity and efficiency were addressed by recognizing that the objective of water management and allocation is an 'optimal benefit to the society'; establishing water use as a temporary right; allowing transfers from one user to another; and, for economic users, promoting the user-pays principle and the polluter-pays principle, including charging for wastewater disposal. All these instruments help to encourage more efficient use of water.

A further step to achieve equity was to 'reserve' the water required to meet basic human needs. Equity in use of water for non-domestic purposes was promoted by designing allocation and financial support mechanisms that would allow formerly excluded communities to gain access to water resources.

For the first time in South African water management, the environment was put centre stage, through the establishment of the environmental reserve. While the implementation of the reserve is necessarily a long-term process, other interventions – notably the approach to pricing water (DWAF, 1999), including the implementation of receiving water quality approaches – have already had significant impacts. These impacts will be reinforced by the introduction of pollution pricing, which was imminent in 2008.

Talking to the users – how water became part of development

Central to South Africa's reform was the explicit understanding that water is an 'indivisible national asset', for which national government is the 'custodian'. Whereas, previously, there had been different categories of water (groundwater, for instance, was considered to be 'private'), the new policy brought the entire water cycle into one management system. This also avoided the fragmentation of water administration between different spheres of government and different sets of rules.

Equally, the emphasis on the river basins as units of management provided a framework for the engagement of stakeholders, particularly since the proposed catchment management agencies were given an explicit mandate to ensure that economic, social and environmental needs were effectively reconciled. By virtue of its water stress, South Africa already has extensive systems of inter-basin connections that require strategic oversight to manage effectively. The requirement for a regularly updated National Water Resource Strategy provides the instrument for such oversight, as well as mandating a structured planning process at a more local level.

These approaches were possible because the legislation reinforced the role of the Department of Water Affairs and Forestry as a strong apex institution to take overall responsibility for water management. The Department's role was strengthened, paradoxically, by the recognition that water had to be managed on a regional (although not river-basin) level if economic, social and environmental needs were to be effectively balanced, and that the voices of users had to be heard in the process. A planning process was therefore developed that took local and regional inputs from different user groups and integrated them into broader national development planning.

Major reform like this takes many years to effect and even longer for its full impact to be apparent. However, if it does not yield early results or worse, if its implementation runs into difficulties, international experience suggests that it will fail. The policies and legislation described here were developed over a four-year period, which allowed extensive consultation with most interest groups.

If it was important to have a national custodian for all water matters, it was just as important to recognize that the water resource managers were not the water users and that user sectors had to be engaged effectively if water management goals were to be achieved.

So the approach to water supply and sanitation services was developed in conjunction with the local government sector. Much of the water services' policy and legislation (see, for instance, Government of South Africa, 1997), notably in relation to planning and the distinction between regulatory oversight and service provision, anticipated and catalysed approaches taken in that sector, although local government did not always accept that the resource management role was not their remit.

Irrigation policy was developed jointly with the Department of Agriculture, even if this did delay the process substantially.

The policy on managing the impacts of mining on water resources was developed in consultation with the Minerals and Energy Department as well as with the Chamber of Mines, which represents the mining industry.

Finally, while water resource management is an environmental management activity in its own right, the policy-makers recognized that the approaches taken had to be consistent with the overarching approach to environmental management. To this end, a great deal of joint work was undertaken with the Department of Environmental Affairs and Tourism so that, notwithstanding the inevitable bureaucratic boundary disputes, overall approaches were coherent and there was cooperation and coordination in their implementation.

Another important channel for integration was through South Africa's national strategic planning system (see Figure 12.4). Because the water resource management process had explicitly addressed the opportunities and constraints created by water in different regions and for different sectors, it was relatively easy to ensure that these were reflected in regional and sector plans as well as in South Africa's National Spatial Development Framework. The value of having the water constraints set out in a document published by the country's presidency, rather than simply by a line department, should

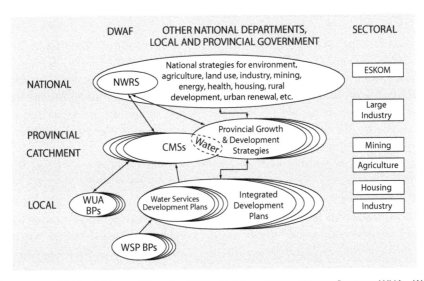

Note: DWAF – Department of Water Affairs and Forestry; CMS – Catchment Management Strategy; WUA – Water User Association; WSP – Water Services Provider; BP – Business Plan; ESKOM – National Electricity Utility

Figure 12.4   Water-related planning in the national planning framework

not be underestimated in any government system. So the inclusion of comments such as 'The dominant pattern of settlement and economic activity in South Africa is largely out of line with water availability ...' in the country's National Spatial Development Perspective is, in itself, an important achievement (Presidency of South Africa, 2003).

The outcomes

The process might have been consultative and inclusive, the approach integrated, but the nagging question remains: what impact did it have? Government and water managers (including the author) claim many successes, although there is also recognition that implementation is proving to be more difficult than anticipated. There are also many critics, some of whom simply write off South Africa's water policy as neo-liberalism inspired by the World Bank (Bond, 2000; Bond and Ngwane, 2001), while others have provided more nuanced accounts of the evolution of policy from civil society

and academic perspectives (Munnik and Phalane, 2004; De Coning, 2006).

Especially since the reforms in water were part of a much larger programme of change, impacts are difficult to isolate and attribute. Nevertheless, even at the level of a whole country, some outcomes can be identified and measured. And since the objective of better water management is to achieve outcomes in all three dimensions – social, economic and environmental – it is to these that we look, in turn to identify what has been achieved in South Africa as a result of good water resource management in the 12 years since a democratic government came to power.

Turning first to the **social dimension**:

- For a start, *good water resources management enabled more than 14 million people to get access to safe water* between 1994 and 2005. Although the resource is very limited and there is significant competition between users, the government's water supply programme was not constrained by conflicts over water resources. In part this is because basic water needs do not usually place a substantial demand on the resource. Even in South Africa, 25 litres per person per day uses barely more than 1 per cent of available water. But even in areas where the resource is overexploited, the priority of basic human needs over all other activities was respected.

- Good management – including secure allocations, clear responsibilities and an explicit pricing policy – made it possible to fund a number of large economic projects using private-sector resources. This is vital in addressing social challenges because it *releases public funding* that would otherwise have been spent on infrastructure to supply industries and well-off domestic consumers who can well afford the costs. Over a ten-year period, users funded more investment in projects such as the Lesotho Highlands, Western Cape's Berg River Dam, the Vaal River Augmentation and others (more than R13 billion or roughly US$2 billion) than was required for the national basic needs programme (around R10 billion, roughly US$1.6 billion).

- Social needs have also been extensively served as part of the process of economic investment. Thus, in a number of privately funded projects, notably in the Lebalelo and Impala Water User Association projects, water allocations and project approvals for economic users were granted on the basis of agreements to reserve some of the *water provided for social purposes.*

In the **economic domain,** key indicators such as water productivity (in terms of contribution to the GDP or jobs per cubic metre) and the real price of water for economic use (an indicator of response to scarcity) have increased (NEDLAC, 2008). And there are many examples where the social and the economic domains overlap and are mutually supportive:

- The construction of the Bivane Dam, intended to *increase water security (and therefore incomes and employment levels) of 16,000 hectares of commercial sugar farms* in the Zululand region was approved on the condition that part of the water it made available must be used to supply 250,000 rural people in the region. It also significantly expanded the irrigated area, creating new opportunities for some of these small-scale users to engage in commercial farming on their own account.

- Similar benefits were made available from a development on the Blyde River: R110 million (US$19 million) was invested by commercial farmers in a new pipeline to reduce water losses between the dam and their mango and citrus orchards; part of the water saved was allocated to *expand irrigation for small-scale farmers although land reform processes have stalled this component of the project.* The pipeline also enabled the farmers to *save energy* by using the pressure from the head of the water in the dam instead of electric pumps to drive their irrigation systems.

- A key outcome of the abolition of riparian water rights has been the expansion of irrigation on land that does not enjoy river frontage. This has encouraged a significant expansion in the production of table grapes for the export market along the Orange River in the Northern Cape. *Agricultural employment in the province has increased significantly* as a result, in contrast to the overall decline at national level.

- Benefits did not just accrue to South Africa. One direct result of water-sharing agreements reached between Mozambique, Swaziland and South Africa over the use of the shared Incomati and Maputo Rivers was to release an *investment of R115 million (US$20 million) in the Lower Usutu Irrigation Project.* This will create an estimated 5,600 jobs in Swaziland, mainly in smallholder farming, secure from the ever-present risk of drought.

- The threat of drought has forced the country to invest in ensuring the water supply, particularly in the dry interior. One response to this was the Lesotho Highlands Water Project, built jointly with neighbouring Lesotho, which stores and transfers water from the Orange River catchment to its Vaal tributary high in the mountains, avoiding the need to pump it up to consumers. While the project has been fiercely attacked by anti-dam activists, it has *boosted Lesotho's GDP by about 15 per cent and created thousands of jobs* in a small country plagued by high levels of unemployment; the project has also provided *water security to the Gauteng region, which is responsible for 60 per cent of South Africa's GDP.*

- Good hydrological analysis, careful planning and targeted investment in infrastructure also yield one key benefit that is too easily taken for granted: the provision of a reliable water supply. Although the climate of South Africa's interior is extremely variable, *no significant water shortages have affected the key economic activities* over the past decade despite some serious droughts. (The one major exception occurred when the construction of a dam to expand Cape Town's water supply was delayed by challenges from environmental groups.) During the same period, while failures in electricity infrastructure and high prices and poor service from the country's transport and telecom utilities were widely blamed for hindering the country's economic progress, the reliability of water supplies was not identified as a constraint.

- Attention to the competing demands for water in the inland areas of the country and awareness of the problems caused by pollution from both domestic and industrial sources have also produced direct benefits. A number of *mining projects in water scarce areas are taking their water supplies from municipal wastewater discharges.* Agreements have been reached in a number of cases between mines and municipalities whereby the mining companies build the sewage treatment works and take the effluent to be used in their processes.

- *Technology for such cooperative treatment has been developed with support from South Africa's user-funded Water Research Commission in response to the challenges identified by the users themselves.* Indeed, one of the strengths of South Africa's reform process has been its ability to turn to the country's university and R&D community for support. This support has often been funded through the Water Research Commission.

Again, these final examples demonstrate how **environmental benefits** flow, almost seamlessly, from the more integrated approach to water management that has been taken in the country:

- Pressure to reduce pollution has seen a number of *major industries based in sensitive inland catchments move towards zero discharge processes.* The measure of success in these cases is the investment that has been made. Thus the steel company Mittal (formerly Iscor) has invested over R500 million (US$83 million) in process improvements to convert its Vanderbijl Park plant on the Vaal River to zero-discharge operation; the paper company SAPPI has spent a comparable amount on new processes to reduce pollution from its Ngodwana plant.

- An explicit focus on water conservation and demand management has also achieved successes. One measure of this is that *domestic water consumption in urban areas grew by only 20 per cent between 2001 and 2006, while the number of people served increased by nearly 50 per cent.* These figures indicate that per capita use of water fell steadily over the period, driven by a variety of factors including a tariff policy explicitly designed to discourage wasteful use of water (NEDLAC, 2008).

- A subtle but important achievement has been to ensure that *water prices reflect the opportunities and constraints* imposed by the natural resource. Thus industrial water prices in inland areas tend to be set at the level of the highest tier of domestic water charges (appropriate given the scarcity and the marginal cost of developing additional supplies). Meanwhile, in coastal areas, prices tend towards the lower and median domestic charges, which are close to the cost of desalination. This is appropriate since desalination will be able to supply industry in coastal areas, and is an indication that the pricing system is sending appropriate signals to water users as to where they should go to expand their thirsty operations (NEDLAC, 2008).

- The environmental gains link back again to social gains. Thus Working for Water, a land management initiative organized as a public works job-creation programme, has cleared over a million hectares of alien vegetation. This work has had a significant impact on the availability of water in some sensitive catchments. However, its most immediate and visible success has been the employment opportunities it has generated – the equivalent of around *12,000 full-time jobs per year* in some of South Africa's poorest rural communities. Thus it has linked water to welfare in a very practical way.

There have been many other important developments. Since to manage you need to measure, an important advance has been to establish a *register of existing actual water use throughout the country*. The registered claims are now being verified, which is a massive task since many users have claimed more than their original entitlements. However, users are funding the management process through their water charges, which they have a strong incentive to pay since, if they do not, they might lose all rights to use water.

One consequence is that water users are now paying more attention to what their neighbours are doing. Following complaints about illegal water use, there were a number of high-profile raids that led to illegal dams being demolished and boreholes being shut down, enforcement activities that contributed to a *growing water awareness* in the country – without which any water reform process is doomed to failure.

This encouraging report should not create the impression that all in the water sector is positive. Some of the agricultural support schemes have

come to a standstill as farmers contest the costs and seek additional government support. In others, planned engagement of small farmers has been halted by disputes over land tenure. The land reform process, intended to promote rural equity, has in many places created conflict and uncertainty, which has reduced the opportunities to use the water that has been made available for social and economic development.

There are concerns about the management of existing water resources infrastructure and about the country's ability to maintain the technical excellence needed for water management in conditions of scarcity. And there are indications that the erosion of institutional capacity is beginning to hinder some of the more ambitious innovations.

On a more mundane and local level, pollution caused by poorly managed sewage treatment works, in part the result of the enthusiastic expansion of services without due attention to managing their wastewaters, is a problem in many areas. The problem is compounded by a general lack of technical capacity in the municipalities to which many powers and functions were decentralized without full understanding of their administrative and financial constraints.

And, although preliminary work is being done, no compulsory licensing process has yet been undertaken in a stressed catchment, where water use has to be reduced in order to meet environmental goals.

Nevertheless, there is a sense that water management is not just staying ahead of economic and social development, but is actively contributing to it. And there is growing understanding of the balance between development and environmental sustainability, although the debates are no less contentious.

Lessons learned

Perhaps the key lesson from the South African experience is that broader political change can

---

Box 12.1   The Limpopo Province grapples with its water constraints

It is sometimes hard to separate the social, economic and environmental outcomes that have been achieved by effective, well-coordinated water resource management that has engaged effectively with its stakeholders. In the Limpopo Province, arguably one of the most water-stressed in the country, water has been at the centre of the province's Growth and Development Strategy discussions. The province's initial priorities were more irrigation, mining and tourism – all of which required more water. After presentations to the provincial cabinet as well as to public planning meetings, it was recognized that water was going to be a major constraint and that, since there would be no more water for agriculture, existing allocations would have to be used more efficiently.

A series of innovative actions have helped enable the province to pursue its goals within the constraints of available water and with active environmental groups, led by the Kruger National Park authorities, campaigning to limit abstractions and reduce pollution:

- Mining companies are cooperating with municipalities in the province to use urban wastewater in mining processes. The mines fund the municipal sewage works and are also responsible for treating their own effluent after the water has been used for a second time.
- A group of competing mining companies joined together to fund and build a R350 million (US$110 million) canal system to transfer water from a surplus area to support expansion of their operations. They included capacity for domestic supplies to approximately 150,000 people along the route.
- Some water was 'rented' from small farmers who used the rental payments to help fund the rehabilitation of irrigation systems that had fallen into disrepair.
- After four years of technical investigations and social and environmental consultations, it was agreed to build the new De Hoop Dam that should start supplying the mines in 2011. It will also support a pumped storage scheme to generate peak electricity, enable minimum flows to be maintained in the Olifants River (which flows through the environmentally important Kruger National Park into neighbouring Mozambique), and provide a secure source for domestic supplies to at least 250,000 people.
- Mozambique agreed to allow construction of the dam on the shared river because there were already agreements in place to protect its interests.
- The communities affected by the various alternative dam sites were actively competing for their site to be chosen. This was in part because, on a previous project (the construction of the Nandoni Dam), levels of compensation were agreed with the rural communities whose dwellings or fields were affected. The terms were sufficiently attractive to encourage other communities to volunteer sites for new dams in their areas.
- Finally, the amount of water available in the province is also being augmented by substantial flows of treated wastewater from the urban areas around Johannesburg, some of which is, in turn, derived from the Lesotho Highlands Water Project.

---

create important opportunities to implement more effective water management arrangements. The water reform is more likely to be successful if it is sequenced to address social objectives at an early stage, since that helps to generate political support for more complex economic and environmental reforms that may take longer to implement.

Another lesson is the importance of having a strong technical institution that could provide a scientific and intellectual foundation from which to understand the resource challenges and respond to them.

The approach to the establishment of new institutions and the allocation reform has been pragmatic – the national water resource strategy programme allows 20 years for it to be completed.

This allows new approaches to be developed and implemented (rather than imposed) at the pace of the people concerned.

Finally, during the transition to democracy, South Africa took the decision not to decentralize the overall function of water resource management. This was based on the experience of other countries with federal constitutions, where the allocation of water management powers and functions to States or Provinces has created major obstacles to the achievement of equitable, efficient and sustainable water use. The notion that water is an indivisible national asset has proved to be an important element of South Africa's water management success so far.

While it may have been the particular conditions of South Africa that led to an attempt to 'do it all' at once rather than move by piecemeal reform, the outcome so far has been positive although many challenges remain. It is not, however, a model that other countries should necessarily follow. Indeed, it was the coincidence of reaching a political threshold at the same time as the water resource constraints became acute that has shaped the South African process. Unless those conditions apply, it would be rash to attempt to replicate it elsewhere.

However, perhaps the final key lesson is that much of the technical foundation for the reforms had been laid before the political transition, which made it possible for them to be implemented. The South African case demonstrates that it is rarely too early to plan for the future.

Note

1. The author served as Director General of South Africa's Department of Water Affairs and Forestry from 1997 to 2005, and thus was intimately involved in much of the process described here.

References

ANC (African National Congress) (1994) *Reconstruction and Development Programme*, African National Congress, Johnannesburg

Bond, P. (2000) *Elite Transition: From Apartheid to Neoliberalism in South Africa*, Pluto Press, London

Bond, P. and Ngwane, T. (2001) *The World Bank and Backward/Forward Influences in Post-Apartheid South Africa*, paper presented to the Centre for Social and Development Studies Project on Donor Funding at the University of Natal, Durban, on 30 October 2001

De Coning, C. (2006) 'Overview of the water policy process in South Africa', *Water Policy*, vol. 8, pp505–528

DWAF (Department of Water Affairs and Forestry) (1994) White Paper on *Water Supply and Sanitation Policy*, Department of Water Affairs and Forestry, Pretoria

DWAF (Department of Water Affairs and Forestry) (1996) *Fundamental Principles and Objectives for a New Water Law in South Africa*, Department of Water Affairs and Forestry, Pretoria

DWAF (Department of Water Affairs and Forestry) (1997) White Paper on *A National Water Policy for South Africa*, Department of Water Affairs and Forestry, Pretoria

DWAF (Department of Water Affairs and Forestry) (1999) 'Establishment of a pricing strategy for water use charges in terms of section 56(1) of the National Water Act, 1998', *Government Gazette*, No. 20615

DWAF (Department of Water Affairs and Forestry) (2004) *National Water Resource Strategy*, first edition, Department of Water Affairs and Forestry, Pretoria

Government of South Africa (1970) *Commission of Enquiry into Water Matters*, Government Printer, Pretoria

Government of South Africa (1996) *Constitution of the Republic of South Africa, Act 108* (as adopted by the Constitutional Assembly), Government Printer, Pretoria

Government of South Africa (1997) *Water Services Act*, No. 108 of 1997

Government of South Africa (1998) *National Water Act*, No. 36 of 1998

Muller, M. (1998) *Water Sector Reform in South Africa – A Midterm Report*, keynote address at the World Bank Water Week in Washington, DC

Muller, M. (2007) 'Parish pump politics: The politics of water supply in South Africa', *Progress in Development Studies*, vol. 7, issue 1, pp33–45

Muller, M. (2008) 'Free basic water – a sustainable instrument for a sustainable future in South Africa', *Environment & Urbanization*, vol. 20, issue 1, pp67–87

Munnik, V. and Phalane, M. (2004) *The SA Water Caucus, Global Justice and Rural Issues: Lessons from and for Civil Society*, Centre for Civil Society, University of Natal, Durban

NEDLAC (National Economic Development & Labour Council) (2008) *Administered Prices Study on Economic Inputs (Ports Rail and Water)*, HSRC (Human Sciences Research Council report) for NEDLAC Trade and Industry Chamber, March, Pretoria

Presidency of South Africa (2003) *National Spatial Development Perspective*, Presidency of South Africa, Pretoria

UNESCO-WWAP (2006) *Water – a shared responsibility. United Nations World Water Development Report 2*, United Nations Educational, Scientific and Cultural Organization (UNESCO), Paris

Part Four – Transnational Level

The approach to be adopted when moving from national to a larger scale is often mirrored by the need for common approaches within countries with federal systems of government, which include the USA, India and Australia. As at other scales, what has to be decided is what issues, actions and decisions should be addressed at which level.

Sometimes this is obvious. The sharing of water within a water basin needs to be agreed at the level of that basin and, if it is shared between two countries, an international agreement will be required. If, as in a federal nation such as Australia, responsibility for water resource management is placed at a state (provincial) level, water sharing has to be treated as an interstate matter, with formal agreements, as has been done in the Murray–Darling Basin.

However, just as with basins which fall entirely within a national jurisdiction, all matters to do with water use in a basin do not necessarily have to be controlled at basin level. Typically, a state or country may apply its own criteria and procedures to govern water that is allocated to it. The exception is with respect to decisions which impact across borders, such as those related to water quality, which will need to be regulated at basin level.

This is demonstrated in the Mekong River where, despite decades of cooperation, individual countries are still making investment decisions (in this case, constructing dams for hydropower production) on a national basis. This can be appropriate if such development is informed by basin-level investigations and the information about investments is shared. Indeed, the major contribution of four decades of cooperation on the Mekong would appear to be the development of a shared information base and a better understanding of issues and interests between the partners.

The limited role for transnational basin organizations is highlighted in the South Africa case, where international agreements have been signed between all basin states, but individual investment projects are promoted by special purpose organizations which include only those countries that are directly involved.

Importantly, water resources management at the transnational scale is not limited to the sharing of water within transboundary water basins. Transnational-scale management issues come into play when countries import or export water resources; Singapore, for example, imports most of its water resources from Malaysia, which has prompted the island nation to consider water as a strategic resource and a key issue of foreign policy. Trade in 'virtual water' – the water used in the production of a good or service – can also be considered as an aspect of water management at the transnational scale, and one that has significant impacts both on global trade policy and on the use and management of water resources at national levels, especially in water-scarce regions.

So at the transnational scale just as at the local level, water management cannot be organized according to a formula. It must be guided by the specificities of each case, which are as complex and diverse at the macro-scale as they are at the local level, and by the overarching principles of equity, environmental sustainability and economic efficiency.

13

Transboundary Cooperation in Action for Integrated Water Resources Management and Development in the Lower Mekong Basin

Hartmut Brühl and Michael Waters[1]

Millions of people in Southeast Asia are bound together by the Lower Mekong River. Given the importance of irrigated agriculture and hydropower generation for national development, the cooperative management of this common resource is a major development challenge for Laos, Cambodia, Viet Nam and Thailand. The Mekong River Commission and its predecessor the Mekong Committee were created to address this challenge.

This case study describes the geographical and political context of the Lower Mekong Basin (LMB), provides details of the governance arrangements for IWRM implementation across the basin, and describes the measures being used within MRC to support decision-making within this IWRM framework.

As the Mekong experience shows, improving management and development of the water resources in a complex transboundary setting requires significant consultation, stakeholder involvement and technical expertise. While some problems require transnational solutions, others demand more local-level solutions and are best managed at national or sub-national scales. But because these activities affect water balances, there need to be links between local-level institutions and decision-making at larger scales. The challenge then is to define institutional arrangements that can coordinate between actors and decision-makers operating at different scales – local, basin, national and transnational.

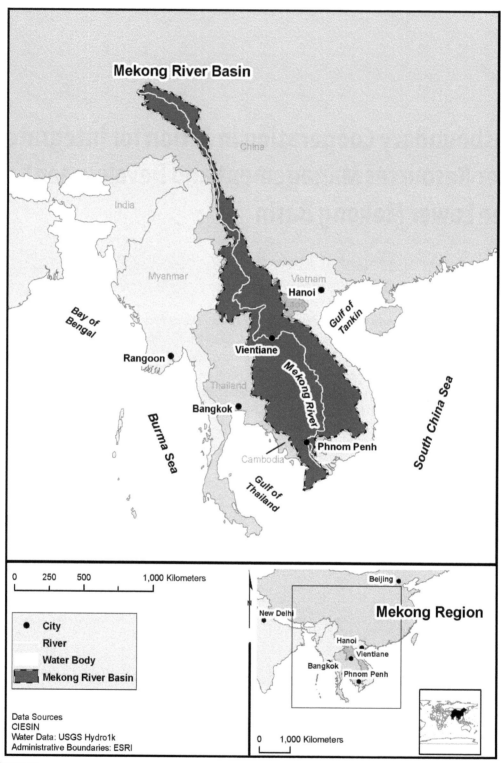

Location map

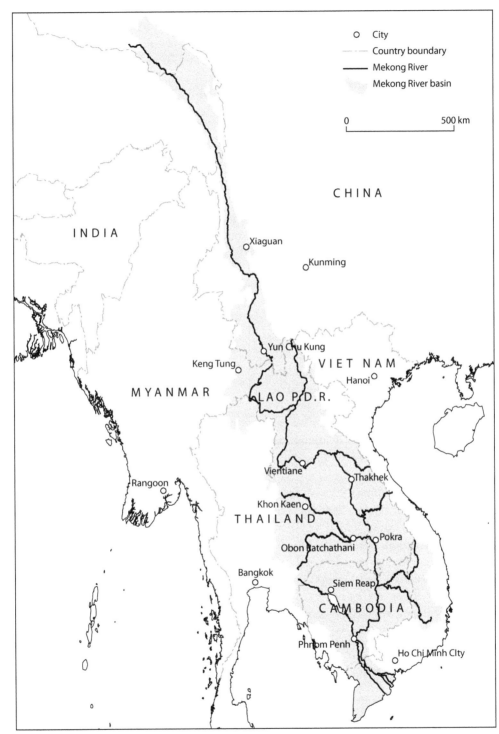

Source: from MRC, 2005b

Figure 13.1   Mekong River Basin showing country catchment areas as percentage of MRB total area

The development context and water challenges

Physical and hydrological characteristics

The Mekong Basin has an area of 795,000 km². The upper basin section in China and Myanmar makes up 24 per cent of this area and contributes approximately 18 per cent of total flows (MRC, 2005a). The LMB includes the upland areas, including the tributaries that drain the highlands of Lao PDR, the Nam Mae Kok watershed in Thailand and the Se Khon-Se San-Sre Pok (Three S) watershed shared by Cambodia, Lao PDR and Viet Nam. Together, these upland areas, which are fertile, forested and have high rainfall, contribute approximately 65 per cent of total flows.

The Songkram, Kam, Mun and Chi watersheds form the major Thai portion of the basin. These tributaries of the Mekong drain the Khorat Plateau, a flat, semi-arid area that contributes approximately 12 per cent of total flows (see Figure 13.1). This area frequently experiences droughts during the dry season.

Further downstream is the Cambodian floodplain and Tonle Sap Great Lake system. Here, annual wet-season flooding causes the Tonle Sap River to reverse its direction, filling the Great Lake upstream from a dry season average depth of 1 m to a wet season average depth of 9 m. The Great Lake has a wet season storage volume of 60 to 70 billion m³, which is equivalent to approximately 12 per cent of the total annual flow volume in the mainstream Mekong River, compared to a dry season storage of less than 1.5 billion m³. As the wet seasons ends, the water stored in the lake is slowly released again through the Tonle Sap River and into the Mekong. This flooding process underpins one of the world's most productive ecosystems. More than 2.5 million people have adapted to living with this annual flood–drought regime.

Finally, downstream of Phnom Penh, the Mekong divides into a complex network of distributaries as it enters and flows through the 55,000 km² delta area in Viet Nam. This channel network has been adapted for irrigation, navigation, drainage and flood prevention. Around 85 per cent of this area is used for agriculture, mostly paddy rice. More than 50 per cent of Viet Nam's national rice production comes from the delta. The delta is mostly tidal, so saline intrusions can become significant in the dry season, leading many farmers to adopt a rotating rice-harvesting, shrimp-raising system.

Box 13.1   Rotating rice and shrimp farming – My Xuyen, Viet Nam

The MRC project to support a sustainable system of rotating salt-water shrimp aquaculture and irrigated rice farming in the Mekong Delta shows a practical, small-scale example of IWRM in action. This MRC-funded project – under the Research Institute for Aquaculture No. 2 in Ho Chi Minh City, Viet Nam – commenced operating in the My Xuyen District, Soc Trang Province of the Mekong Delta in Viet Nam in 2003, and is documented fully by the MRC (2006d).

The rotating system promotes economic productivity by ensuring sustainable incomes based on rice production, a relatively stable commodity, while allowing more risky but lucrative shrimp production during the dry season, when waters in the coastal parts of the Mekong Delta are relatively brackish and are therefore not suitable for rice farming.

Small-scale IWRM is vital to the success of such projects. This is seen most dramatically in that the coordination of water movements at district level is necessary to ensure favourable outcomes for all. For example, if a farmer wants to pump saline water into his pond from the delta too early in the dry season for shrimp production, this may cause salinization of surrounding rice crops and consequent failure. Similarly, clashes can occur if some farmers drain polluted ponds at a time when other farmers wish to fill their ponds.

Farmers' cooperative groups have now been established that incorporate hatcheries and buy seed from other areas. Cooperative members pay monthly membership fees to support monitoring of water quality and use, a hatchery and seed buying, and to build cash reserves to cover losses that may occur if crop failures should occur. Long-term sustainability means that farmers must carefully consider their shrimp stocking rates. Stocking too many shrimp causes shrimp faeces and uneaten feed to build up in the ponds, which over time means crop yields diminish. The favoured model is for one rice crop and one shrimp harvest per year, since more intensive systems appear to be unsustainable in the long term.

The project includes a strong capacity building element for farmers, since maintaining healthy shrimp stock alternately with productive rice yields, ensuring water quality is maintained and considering the varying movements due to tide and river flows of fresh and saline water are highly technically challenging.

Overview of social issues

Many of the primary social issues in the LMB are documented in the *Social Atlas of the Lower Mekong Basin* (Hook, Novak and Johnston, 2003). Significant issues within the basin identified in the *Social Atlas* include the following:

- Approximately 60 million people live in the LMB. Population density is highest in Phnom Penh, then in the Mekong Delta (Viet Nam) and lowest in rural Cambodia and Lao PDR. Approximately 80 per cent of the population live in rural areas and rely heavily on natural resources for their livelihoods.
- Poverty is a critical issue across the basin. The proportion of the population living below the poverty line exceeds 50 per cent in parts of each country within the basin. The full details of poverty distribution are quite complex. For example, some remote areas in northern Lao PDR have poverty rates above 50 per cent but quite low population densities. By contrast, in areas around the Tonle Sap Great Lake, in the Mekong Delta and central Viet Nam highlands, poverty rates are not as severe but population densities are higher, meaning the overall number of poor people in these areas may be the same or higher than in the remote but less inhabited areas of northern Lao PDR.
- There are serious development-related health issues in parts of the basin. Within Lao PDR and Cambodia, infant mortality rates are as high as 12.5 per cent of live births (UNDP, 2002). Child malnutrition and malaria issues are also quite significant. HIV prevalence within the basin ranges from 0.1 to 2.8 per cent, with the lowest rates in Lao PDR and highest in Cambodia.
- Most of the population in Thailand and Viet Nam have access to safe water and sanitation, although some areas of the Mekong Delta are problematic. Currently, rural populations in Cambodia and Lao PDR have poor access to water and sanitation.
- Unemployment rates within the basin tend to be low because the generally rural-based populations of the basin largely practise subsistence farming, which accounts for almost all available labour.

Approach to water resources management in the Mekong Basin

Origin of the approach

In 1957, when the Committee for Coordination of Investigations of the Lower Mekong Basin (the Mekong Committee), under the guidance of the United Nations' Economic Commission for Asia and the Far East (UN ECAFE, later ESCAP), was established, the riparian countries Laos, Cambodia and South Vietnam had been independent for only a few years. This was a time of great political tension with Viet Nam divided and war imminent. Economic development was low and all four countries relied on the water resources of the Mekong River as their lifeline.

Water was and is crucial for the development of all four countries of the Lower Mekong Basin (LMB) in many aspects, the single most important use being irrigated agriculture followed by power generation. In addition, due to bad or nonexistent roads in many areas, water transport was and continues to be crucial for marketing agricultural products. Therefore, the idea behind the Mekong Committee was to foster economic and social development of the riparian countries through coordinated development of what was thought of as one of the world's great 'untamed rivers', a project that had clear political as well as social objectives. Although based on the model of the Tennessee Valley Authority (TVA), this project was groundbreaking because it was the first attempt to take on such encompassing responsibilities for investigating, planning, financing, constructing and managing projects on an international river and its tributaries. The reality, however, turned out to be far more modest; the new basin organization confined itself to collecting data, investigating, studying and planning projects, as well as assisting in fundraising.

Although the war in Viet Nam had ended, conflict in the region continued through the 1970s, principally in Cambodia, resulting first in an interruption of the Mekong Committee sessions and later, in 1977, in the formation of the 'Interim Mekong Committee' composed of Lao PDR, Thailand and Viet Nam.

The new Mekong River Commission

After political stabilization in the early 1990s and the re-admittance of Cambodia, the Mekong Commission was transformed by the 1995 'Agreement on the Cooperation for the Sustainable Development of the Mekong River Basin', in which the four LMB countries agreed to 'cooperate in a constructive and mutually beneficial manner for sustainable development, utilization, conservation and management of the Mekong River water and related resources'. The Agreement established the Mekong River Commission (MRC) and gave full management responsibility for the Commission to a Council of Ministers of member countries.

The full institutional framework for integrated water resources management across the LMB is complex. At LMB level, there are at least 25 international and regional conventions, treaties, protocols and declarations that are relevant to water and land management (MRC, 2005c). Current trends suggest that, for much of the LMB, IWRM will be implemented across the basin at three levels:

- at sub-basin level, where community-centred watershed management committees are being developed;
- nationally, where administration, planning and legislative responsibilities for water resources in the countries are generally spread over a number of different ministries in a sector-driven manner that differs in each country; and
- at the basin-wide level as described in this case study.

Thus far, the experience of the MRC and its predecessor organization suggests that implementation of IWRM on the ground will be gradual in the member countries, because important changes in mindset are required.

Besides MRC, at least 40 organizations within the basin, regionally and internationally, have roles to play in IWRM in the LMB. These include organizations such as UNEP, UNDP, UN-ESCAP and FAO, non-government organizations (NGOs) such as IUCN and WWF, and Universities. Strategic partnerships with other regional initiatives are vitally important to IWRM implementation in the Mekong, particularly the Association of South East Asian Nations (ASEAN), the Asia Development

Bank's Greater Mekong Subregion Economic Cooperation Programme, and the World Bank-ADB Mekong Water Resources Assistance Programme (MWRAP).

Related water resources issues for the Mekong

Particular water resources sectors that are important for the LMB are fisheries, flood management, navigation, agriculture/irrigation, hydropower and the environment.

Fisheries

The aquatic ecology of the LMB supports the most abundant inland freshwater flora and fauna in the world after the Amazon. In the LMB, over 1,200 fish species are recorded and an estimated 2 million tonnes of fish and other aquatic animals are consumed annually (MRC, 2005b). Many rural communities are heavily dependent on fisheries for their livelihoods and basic subsistence – making fishing critical to food security in the LMB. The annual total value of the Mekong fisheries is estimated at approximately US$1.4 billion.

A good example of the historical interconnectedness between social, economic and environmental benefits that the water resources of the Mekong give to the people of the region is the dai fisheries of the Tonle Sap River in Cambodia. Falling water levels in the dry season trigger the migration of large numbers of fish of the *Henicorhynchus* genus, known in the Khmer language as *try riel*, from the Tonle Sap Great Lake through the Tonle Sap River, to the Mekong mainstream. This migration causes the 'dai' fishing nets of Cambodia to swell. This catch alone supplies approximately 10,000–20,000 tonnes of fish, a vital food resource to the people of Cambodia worth up to US$ 6.2 million annually (Halls, Sopha and Pengby, 2007).

Tonle Sap River dai catch values are an important indicator of the ecological health of the Tonle Sap Great Lake, which is a particularly critical water resource for the Mekong. High catch values are a good sign that economic, social and environmental benefits can be sustained in the Tonle Sap Great Lake, and the Mekong more widely.

Flood management

The yearly floods in the Lower Mekong Basin (LMB) bring great benefits. They are linked to important catches of fish; they flush pollutants away; they refill the wetlands that provide shelter, food and nurseries to the region's wide variety of birds, mammals, reptiles, amphibians, fish and invertebrates. The annual high and low flows of the river support traditional flood recession agriculture and deposit sediments that provide silt and clay for brick-making, and sand and gravel for traditional construction materials.

However, flooding can on occasion bring great hardship. When 500 km^3 of water flowed down the river between 1 June and 30 November 2000, the most severe floods in the Mekong in more than 70 years resulted. This amount of water, large enough to flood an area equivalent in size of India to a depth of more than 0.15 m, caused flooding across 22 of the 24 provinces in Cambodia, with over 3.4 million people affected and almost 400,000 evacuated from their homes (MRC, 2001). In Lao PDR approximately 400,000 people were seriously affected, in Thailand 2.8 million and in Viet Nam 5 million. Across the whole LMB there was a death toll of over 800 people (mostly children) and almost US$500 million of damage. With climate change expected to aggravate such extreme events and the populations of vulnerable areas growing, the natural cycle can impose huge social and economic costs.

Because the flooding has both positive and negative effects for the population and, more generally, because of the wide range of issues in the LMB – including initiatives that will inevitably damp the flood cycles, such as expanding irrigation and hydropower generation – an integrated approach to water resources management and development is highly necessary.

Navigation

The Mekong River is navigable in two sections. The lower section reaches from the Mekong mouth in Viet Nam to Cambodia (including the Tonle Sap Lake) below the Khone Falls. This stretch supports relatively large vessels of up to 5,000 deadweight tonnes (DWT) as far as Phnom Penh (Geerinck et al, 2004). During the dry season, the upper section, which stretches from Khone Falls though Lao PDR, Thailand and Myanmar to Jing Hong in China, can support only 15 DWT along some parts of the river.

Agriculture

Over 40 per cent of the land area of the LMB is used for agriculture, particularly rice farming, which overwhelmingly dominates livelihood patterns of the people living there. Irrigated agriculture accounts for approximately 42 km^3 of withdrawals from the mainstream and tributaries. These withdrawals account for 9 per cent of the total flow in the river, making agriculture the largest water use sector in the LMB (Young, 2005). About 26 km^3 of this withdrawal takes place in the Mekong Delta in Viet Nam.

Hydropower

Hydropower is the major natural resource available for power production in the LMB. While the renewable nature of hydropower is a particular advantage, environmental and social impacts cannot be ignored. Across Lao PDR, Cambodia and Viet Nam, over 60 sites on the tributaries and mainstream of the Mekong have been identified for potential hydropower developments. These sites have a total potential capacity of 30,000 MW. Of this 13,000 MW are on the mainstream, 13,000 MW in Lao tributaries, 2,200 MW in Cambodian tributaries, and 2,000 MW on Viet Nam tributaries. Thailand does not propose to develop any further hydropower schemes in the LMB; however, Thailand is generally considered supportive of hydropower development in the basin as a power consumer.

To date, 12 medium to large projects (over 10 MW) representing about 7 per cent of the potential in the LMB have been built. These investments have significant indirect and direct economic value to the countries; for example, most of Lao PDR's foreign currency comes from the export of hydroenergy. In spite of all its advantages, hydropower development in the basin also has its constraints and is debated on grounds that are social, environmental and to a lesser extent economic. Therefore, it is difficult to foresee how much of the potential hydropower capacity will eventually be developed.

Environment

Many different ecosystem types and individual species are present in the Lower Mekong Basin, including aquatic ecosystem types and biota that are internationally unique. Numerous conservation areas have been declared, including many sites recognized internationally through listing as World Heritage sites or wetlands of international importance under the Ramsar convention.

Linkages between environmental issues and social development issues are extremely close in the LMB as many communities, particularly the rural poor, are heavily dependent for their livelihoods on the rich natural resources of the Mekong River and its tributaries, including some of the most environmentally significant areas in the basin. Consequently, addressing social and economic development in the LMB generally also means addressing environmental conservation in a forthright manner.

Water quality in the LMB is comparable to that of other major river systems internationally, in that the chemical composition measured at Kratie in Cambodia (upstream of the Cambodian Floodplain) is very close to the mean of compositions for world rivers (Wetzel, 1983). At a basin-wide level there is little direct evidence of declining water quality (Hien, 2003). Nevertheless, the potential for water-quality degradation in the future points to the need for close attention to monitoring and assessment of water quality so that informed decisions can be made on this and other environmental issues.

Enabling instruments

The Mekong Agreement

In order to deal with water resources issues such as those described above in an effective way, an integrated approach to water resources management at national and/or basin level is widely recognized as desirable. The approach to transboundary management of water and related resources that has been adopted by the LMB countries is defined by the Mekong Agreement.

Water resources cooperation in the Mekong Basin commenced in 1957 under the guidance and support of the United Nations (UN). Through enormous political challenges, this cooperation continued under UN patronage until 1995 when the Governments of Cambodia, Lao PDR, Thailand and Viet Nam signed the 'Agreement on the Cooperation for the Sustainable Development of the Mekong River Basin' (the Mekong Agreement). Although the upper basin countries, China and Myanmar, are not signatories to the Mekong Agreement, they are recognized as 'dialogue partners'. The Mekong Agreement has a provision for them to become full members in the future at the time of their choosing.

The Mekong Agreement confirms the LMB countries agree to cooperate for 'the management of water and related resources of the Mekong River Basin' (MRC, 1995), through the Mekong River Commission (MRC). As such, MRC's focus is the sustainable management and development of the Mekong Basin's water resources for socially appropriate economic development, while seeking to protect the unique environmental features of the basin.

'In order to promote a positive harmonious, conflict free approach to water resources cooperation', the Mekong Agreement places its emphasis on cooperation and agreeing on common interests, while aspects of enforcement are deliberately de-emphasized. Equally, it is not in the regional tradition to address controversial issues openly, let alone conflicts of interest. On the one hand, this tradition has slowed progress; on the other hand, it may be the main reason a Lower Mekong Basin organization has managed to survive for half a century.

The Mekong Agreement clearly expresses an intention to drive forward economic development in balance with social equity and environmental protection. This reinforces another tenet of the Agreement: that economic development must proceed to ensure social development, equity and poverty reduction, while maintaining important environmental values.

According to official protocols, as well as in a pragmatic sense, planning and decision-making for basin-wide water resources issues in the LMB is a shared responsibility: the Joint Committee and MRC Secretariat (MRCS) play their roles as mandated by the 1995 Mekong Agreement for basin-wide data acquisition and planning; the National Mekong Committees (NMC) play the role of country coordination, a role that is not specified under the 1995 Mekong Agreement; and, finally, the national government line agencies are generally responsible for implementation, since the Mekong Agreement is explicitly a cooperative agreement on the basis of sovereign and territorial integrity (MRC, 1995).

The Basin Development Plan (BDP)

The Basin Development Plan (BDP) is the principal integrating tool for the LMB at a regional level. The BDP enables analyses of scenarios and the identification/prioritizing of projects for both infrastructural and non-structural developments. From 2001 to 2006, under the supervision of the Joint Committee, the MRCS led a participatory process for developing the BDP with the National Mekong Committees coordinating stakeholder involvement in each country. The process was made up of the following steps (MRC, 2006b):

- comprehensive studies and analyses of water allocation opportunities and needs at sub-area and regional levels;
- scenario analyses of development opportunities, options and constraints at a whole-basin level;
- formulating regional development strategies that fit into scenario analyses and address sub-area and regional opportunities and needs;
- identification of development projects that fall under the regional development strategies;
- establishing a database of identified projects; and
- screening identified projects in order to prepare a list of priority projects.

The approach to IWRM under the BDP is explained in more detail in the section on pages 199–200.

The basin-wide governance approach – water utilization procedures

Article 26 of the Mekong Agreement defines how (and what) rules and procedures the countries may agree on for water utilization in a basin-wide context. These agreements are generally recognized as critical to ensuring that integrated management and development is practised effectively in the basin (Chenoweth et al, 2001). Procedures developed to date cover:

- Data Information Exchange and Sharing (agreed in 2001) – specifies the steps the four countries and the MRC Secretariat (MRCS) are willing to take to operationalize data and information exchange, to make basic data and information available for public access, and to promote understanding and cooperation among the countries for sustainable development.

- Water Use Monitoring (agreed in 2003) – specifies the steps the countries are willing to take in monitoring water uses and intra-basin transfers.
- Notification, Prior Consultation and Agreement (agreed in 2003) – specifies the steps the countries are willing to take to share information on intra- and inter-basin diversions on tributaries and the mainstream of the river.
- Maintenance of Flows in the Mainstream (agreed in 2006) – specifies the steps the countries are willing to take to ensure that acceptable minimum monthly flows are maintained, that the natural reversal of the Tonle Sap takes place, and to prevent average daily peak flows from exceeding what occurs naturally.
- Maintenance of Water Quality (currently under consultation) – specifies the steps the countries are willing to take to ensure the quality of water in the LMB is suitable for beneficial use.

The development of these procedures and associated guidelines has involved a number of technical challenges, which have also led to the initiation of a Decision Support Framework (DSF) and an Integrated Basin Flow Management (IBFM).

Addressing the three Es

As for any river basin, it is critical for water management in the LMB to address economic, social and environmental implications of developments and to determine integrated solutions for sustainable development of the river's water and related resources (MRC, 2006a).

Particularly important in this regard is the development of tools for holistically evaluating triple bottom-line effects of possible development scenarios. In the LMB, the development of such tools is currently under way through the Integrated Basin Flow Management (IBFM) project. IBFM is being implemented under the BDP to assess the ways in which basin development scenarios could cause changes in flows of the river and the economic, environmental and socio-economic changes that would result.

The intention is to enable decision-makers, planners and managers from the LMB countries to assess water resource development scenarios from a sustainability perspective. The IBFM approach borrows significantly from international best practice in Environmental Flows Methodology, but additionally includes economic and social analyses of scenarios for basin development in order to address sustainable water resources allocations in an integrated way.

As such the IBFM includes:

- assessment of the macroeconomic impacts of scenarios;
- use of MRC's hydrologic models (known as the Decision Support Framework, DSF) to determine flow changes due to scenarios; and
- evaluation of socioeconomic and environmental impacts of these flow changes.

While tools for assessing macroeconomics and hydrology are readily available, determining the socioeconomic and environmental impacts of flow changes is a particular challenge, since it involves determining the relationships between a whole host of parameters. This model is still being developed in the form of the 'Mekong Method for Flow Assessment'.

However, it has to be borne in mind that model results are only as good as the inputs and the algorithms and can only serve as additional information within the decision process. Experience in the Mekong Basin, as in other basins, suggests that eventually decisions will not be based on models, but on consultations and negotiations between actors, which, at best, may be informed by data from the models.

What is integrated and how?

The current vision of transboundary IWRM implementation across the LMB is described by MRC (2006a), whereby an IWRM approach at a transboundary/basin-wide level is to be taken, promoting and supporting the uptake of IWRM principles within the member states by:

- promoting sustainable and coordinated development;
- providing a regional cooperation framework;
- knowledge management and capacity development; and
- environmental management through monitoring and protection.

Aspects of integration that are particularly important are an integrated approach to planning of water resources management and development for the basin, and the monitoring and analysis of impacts.

Integrated planning under the Basin Development Plan

BDP implementation during Phase 1 particularly sought a high level of participation and ownership by stakeholders. Within the governments of the four countries, more than 200 line agencies across the four countries have been involved in the BDP process and consultations at national and provincial level (MRC, 2006c).

The comprehensive studies that have been conducted under the BDP have confirmed a number of issues that will need to be addressed as priorities in the future. These include (MRC, 2006c):

- economic development for poverty alleviation;
- social development and equity to ensure fair allocation of water resources and services across different economic and social groups: reduce conflict, promote socially sustainable development;

- regional cooperation to integrate and coordinate water resources development and management between countries: optimize benefits from joint resources, minimize risk of water-related conflicts;
- governance to ensure open transparent and accountable institutions and regulatory frameworks at all levels;
- protection of the environmental, natural resources, aquatic life and conditions and ecological balance from harmful developments; and
- the prevention, mitigation and minimizing of suffering and economic loss due to climate variability.

To implement the BDP under Phase 2, which commenced in 2007, these issues will need to be addressed in a practical manner. The MRC Strategic Plan 2006–2010 (MRC, 2006a) proposes that the BDP establish an IWRM Rolling Plan to identify, categorize and prioritize projects ranging from local to basin-wide scales in order to deliver practical sustainable benefits to the people of the LMB.

Data collection, monitoring and analytical tools

Data, information and knowledge are indispensable to ensure that water resources management and development are planned and coordinated on the basis of environmentally, socially and economically sound principles. Such data collection contributes first in establishing baseline conditions for the present case and second in monitoring changes that take place over time in order to evaluate effects of developments and impacts.

Rainfall, water-level readings and discharge estimates across much of the basin are available from the early 20th century onwards, although understandably gaps occur in several places due to civil conflicts and other reasons. Provisions to improve the timeliness of hydrological data reporting are particularly important for flood management and mitigation to allow emergency warnings that are clear, timely and accurate.

With these considerations, basin-wide Water Quality Monitoring (WQM) has been ongoing since 1985, initially under the interim Mekong Committee, prior to the 1995 Agreement, and then under MRC since that time. Over that period approximately 90 stations, 55 primary and 35 secondary, throughout the basin have been monitored for a suite of 22 physical and chemical water quality parameters.[2]

Recent developments in water quality monitoring include the establishment of a biomonitoring programme (Davison et al, 2006) and consideration for the development of broad Water Quality Indices (WQIs). The WQIs integrate across physico-chemical parameters to look at the impacts of human activities on water quality and to consider the suitability of water for aquatic life and agricultural use. Both biomonitoring and WQI assessments verify that water quality at a basin-wide scale is generally good (Davison et al, 2006; MRC, 2006).

Other data holdings that are significant to effective IWRM implementation include:

- administration – district and provincial boundaries;
- transportation in terms of river ports and road networks within the LMB;
- cultural data on the distribution of ethnic groups throughout the basin;
- climatic data, particularly rainfall;
- hydrological data including water levels, discharge estimates;
- inundation areas on flood plains for wet-season periods;
- protected areas and Ramsar listed sites in the basin;
- topography – a digital elevation model of the basin and bathymetry of the river; and
- remotely sensed/satellite images from third parties such as LANDSAT, MODIS, SPOT, IKONOS, ASTER, RADARSAT, JERS SAR, ENVISAT ASAR and the Space Shuttle.

A suite of analytical tools have been developed under the MRC and the Integrated Basin Flow Management (IBFM) project, including the Decision Support Framework (DSF), which includes visualization tools that allow modelling results to be interpreted and presented graphically, and the Mekong Method for Flow Assessment, still in the development stage, which will allow better prediction of socioeconomic and environmental impacts of flow changes due to potential human interventions or to externally imposed changes such as Climate Change (MRC, 2005a).

The flow response relationships are held in a database known as the IBFM interactive predictive tool. These relationships can become very complex as changes in some environmental and socioeconomic benefits and impacts may also lead to changes in other parameters. Across the six hydrological zones and four bio-hydrological seasons of the basin, the number of parameters is very large. Many of the relationships are currently imperfectly known due to a lack of data, so research is currently under way to fill in these gaps.

Outcomes and impacts

A snapshot of some recent outcomes is given in Table 13.1, which demonstrates significant achievements in data acquisition, planning and capacity development. Activities in these areas encompass overall governance, economic, environmental and social fields, across a variety of sectors. They are unavoidably largely sectoral in their character and need to be utilized in common and over different scales in order to be really 'integrated'.

Compared to many other river basins in this area and elsewhere the data availability in terms of quantity, quality and time series is remarkably good – surely an important achievement.

Remaining challenges

However, as said before, country coordination and project implementation remain the jurisdiction of the national governments and their line agencies. And under the present set of agreements, countries cannot be forced to present intended projects contained within their territory to the MRC for approval, even if they are likely to have impacts on water availability or quality in neighbouring countries.

The difficulties can be seen in the approach to dam projects since 2007 in Cambodia, Laos and Viet Nam, where, at present, the MRC, as well as regional and international institutions such as the ADB and the World Bank, are being bypassed and replaced by bilateral agreements. This shows that all the good will for sectoral integration and regional cooperation that has been expressed in drafting conventions still needs to be made effective for real policy. It is hoped that the resilience shown by the MRC and its predecessors will in the end prevail, and that the countries will plan and implement their new schemes following the rules and recommended procedures of the agreed Basin Development Plan, which may become a litmus test for the MRC.

It will be up to the MRC to demonstrate to the national governments the benefits that can be derived from joint and integrated development of the basin and its resources. The free and un-manipulated access to data and information across levels and scales (Lebel et al, 2005) by all stakeholders is a precondition for any true reconciliation of interests – and the MRC should play a key role in that respect. The increasing interconnectedness of the economies of the Mekong region suggests that new opportunities for cooperation in development of water and related resources may be created, despite the huge differences in political systems and rates of economic growth.

Table 13.1   *Recent achievements in the Mekong Basin*

Basin development planning	Development of routines for identification, scoping and screening of projects; Promotion and facilitation of priority projects and programmes; and Reports for basin-wide IWRM implementation including: State of the Basin Report, MRC Social Atlas, MRC People and Environment Atlas, BDP Planning Atlas.
Fisheries	Strengthening community fisheries and fisheries co-management; Addressing gender issues in fisheries management; Improved protection and conservation measures for endangered species; Improved understanding of fish migration and triggers; Successes in aquaculture of indigenous species: domestication and artificial reproduction; and Capture fisheries assessments.
Hydrology and flood management	MRC Report: 'An Overview of the Hydrology of the Mekong'; Establishment of the MRC Regional Flood Management Centre in Phnom Penh; and Annual Flood Monitoring Reporting from 2005 onwards.
Navigation	Navigation strategy for the LMB; Legal study for the development and implementation of Freedom of Navigation on the Mekong River prepared, including new Draft Protocol between Cambodia and Viet Nam; Management Information System (MIS) for the Port of Phnom Penh, Phase 1; and Launching of a 24-hour navigation aid system for the Mekong from the Viet Nam border to Phnom Penh in Cambodia.
Environment	Habitat mapping and valuation of significant wetlands across the LMB; A functional water-quality monitoring network now containing over 20 years of data; and Establishment of an Ecological Health Monitoring Network.
Capacity building	Fisheries co-management training for around 120 mid- to senior-level management staff and more than 500 users from community groups. Participation of 1,600 government staff and about 4,000 users in management of river and reservoir training activities; Regular training workshops in watershed management; Training on MRC's hydrometeorological data; Short courses on Transboundary Ecological Risk Assessments; Workshops on basin-wide water quality monitoring and data analysis; and Technical workshops on satellite rainfall estimation, radio and internet use in flood forecasting and flood management.
Governance – procedural	Through the water utilization programme at MRC, the four LMB countries have agreed on procedures for data information exchange and sharing, water use monitoring, notification, prior consultation and agreement, and maintenance of flows in the mainstream, with procedures for maintenance of water quality currently under consultation.

Lessons learned

From the establishment of the 1995 Mekong Agreement, management of water resources has developed under a consensus-based IWRM framework across the four countries of the Lower Mekong Basin (LMB).

The experiences in the LMB under the Mekong Agreement show that it is possible to foster sustainable management and development of water resources under a consensus-based IWRM framework in a complex transboundary setting. A combination of significant consultation, stakeholder involvement, high levels of technical expertise and application have been required to make this a reality.

However, it also shows how difficult it is to go from principles that have been jointly accepted to real actions when there are massive national and economic interests at stake. It also highlights the need for clear demarcation of responsibilities so that IWRM approaches are not limited to data collection, but are integral to planning and project implementation even if that is mainly done under national competence. Finally it shows that regional IWRM never comes easily by itself but needs to continuously be fought for, even in an institution with a tradition of half a century.

Notes

1. Michael Waters served as Senior Environment Specialist at the Mekong River Commission Secretariat from June 2006 to December 2007.
2. Sites in Cambodia only came into the programme from 1993, with 21 stations being monitored.

References

Chenoweth, J., Malano, H. and Bird. J. (2001) 'Integrated River Basin Management in the Multijurisdictional River Basins: The Case of the Mekong River Basin', *Water Resources Development*, vol. 17, issue 3, pp365–377

Davison, S.P., Kunpradid, T., Peerapornisal, Y., Nguyen, T.M.L., Pathomthong, B., Vongsombath, C. and Pham, A.D. (2006) *Biomonitoring of the Lower Mekong and Selected Tributaries*, MRC (Mekong River Commission) Technical Paper No. 13, Mekong River Commission, Vientiane, Lao PDR

Geerinck, L. et al (2004) *The People's Highway – past, present and future transport on the Mekong system*, Mekong Development Series, No. 3, April 2004, Mekong River Commission, Phnom Penh

Halls, A.S., Sopha, L. and Pengby, N. (2007) 'Landings from Tonle Sap dai fishery in 2006–07 above the 12-year average', *Catch and Culture*, vol. 13, issue 1, pp7–11

Hien T.T.T. (2003) *Water Quality of the Lower Mekong Basin*, conference proceedings of the 2nd Asian Pacific International Conference on Pollutants Analysis and Control held in HCM City, Viet Nam, from 1 to 3 December 2003

Hook, J., Novak, S. and Johnston, R. (2003) *Social Atlas of the Lower Mekong Basin*, Mekong River Commission, Phnom Penh

Lebel, L., Garden, P. and Imamura, M. (2005) 'The politics of scale, position, and place in the governance of water resources in the Mekong region', *Ecology and Society* vol. 10, issue 2, p18, available online at http//www.ecologyandsociety.org/vol10/iss2/art18

MRC (Mekong River Commission) (1995) *Agreement on the Cooperation for the Sustainable Development of the Mekong River Basin*, Mekong River Commission, Chiang Rai, Thailand

MRC (Mekong River Commission) (2001) *MRC Strategy on Flood Management and Mitigation*, Mekong River Commission, Phnom Penh

MRC (Mekong River Commission) (2005a) *Hydrology of the Mekong Basin*, Mekong River Commission, Vientiane, Lao PDR

MRC (Mekong River Commission) (2005b) *Lower Mekong Basin Fisheries Annual Report, April 2004 to March 2005*, Mekong River Commission, Vientiane, Lao PDR

MRC (Mekong River Commission) (2005c) *Strategic Directions for Integrated Water Resources Management in the Lower Mekong Basin*, Mekong River Commission, Vientiane, Lao PDR

MRC (Mekong River Commission) (2006a) *The MRC Strategic Plan 2006–2010*, Mekong River Commission, Vientiane, Lao PDR

MRC (Mekong River Commission) (2006b) *The MRC Basin Development Plan, The BDP Planning Process*, Mekong River Commission, Vientiane, Lao PDR

MRC (Mekong River Commission) (2006c) *The MRC Basin Development Plan*, completion Report for Phase 1 for the Mekong River Commission, Mekong River Commission, Vientiane, Lao PDR

MRC (Mekong River Commission) (2006d) 'Rice returns as delta farmers reconsider stampede into shrimp', *Catch and Culture*, vol. 12, issue 3, available online at http//www.mrcmekong.org/programmes/fisheries/cc_vol12_3Dec06.htm#3

UNDP (United Nations Development Programme) (2002) *Human Development Report, 2001*, Oxford University Press, New York

Wetzel, R.G. (1983) *Limnology*, 2nd edition, Saunders, Philadelphia, PA, USA

Young, A. (2005) *Irrigation Water Use Assessment*, consultant's report to the Programme to Demonstrate the Multi-Functionality of Paddy Fields over the Mekong River Basin, Mekong River Commission, Vientiane, Lao PDR

14

Conclusions: Lessons Learned and Final Reflections

Roberto Lenton and Mike Muller

The previous 12 chapters have described a broad array of circumstances in which water is managed by different communities with different objectives and at different scales. This chapter seeks to draw some common lessons from this diverse experience and to use them to contextualize some of the current debates about what constitutes better water management. The chapter also offers some final reflections on the IWRM approach itself, looking both backwards at the evolution of the concept and forwards towards the challenges that lie ahead, in particular the complexities of applying the approach in practice amid the ongoing challenges of participation and adaptation.

Drawing lessons from diverse experience

From the 160 farming families of Angas Bremer and the villagers of Sukhomajri to the teeming populations of the Lerma–Chapala and Mekong River basins, one critical message comes through from the cases that have been described: effectively managing water requires sustained collective effort and engagement if it is to be successful.

However, successful outcomes are not necessarily those where there is general consensus or everyone's needs are met. The current acrimonious dispute between conservationists and other water users on the USA's Snake River about whether and how to reverse the decline of the local salmon population is a good example and suggests that a structured standoff may also be a successful outcome. The legal framework that facilitates engagement between widely different positions (that would either remove dams or keep generating hydropower), and offers an independent determination of a balance between them, is in itself an important achievement. So better water management also needs robust, competent and trusted institutions.

Given more time, the parties may reach a more definitive outcome, as occurred in Japan, where bitter conflicts over the management of Lake Biwa have been resolved and an apparently acceptable equilibrium achieved between water supply to downstream users, the industrial and agricultural activities (and their pollution) of upstream users, local tourism and environmental protection.

Often, however, neither the conflicts of interest nor the nature of a successful outcome are as well defined or as clearly related to water management as they proved to be in the Lake Biwa case. Indeed, as described in the introductory chapter, the outcomes of better water management are often found outside the water sector. So in Mali, the success of water management in the Office du Niger irrigation scheme is measured by the fact that farm incomes and other livelihoods have improved as a result of a series of institutional reforms, some of which related to water only peripherally. That water

is now better managed and more efficiently used with greater social benefit is a welcome by-product, but was not seen locally as the primary goal.

In countries such as Chile and South Africa, the outcomes of water reforms remain controversial. There are concerns that too many benefits may have flowed to particular sector interests (hydropower generators in Chile) or reached too few in others (subsistence farmers in South Africa). In the absence of a clear methodology supported by more detailed information, it remains difficult to determine whether the adoption of IWRM approaches has contributed to environmental sustainability or whether it has helped or undermined the resilience of livelihoods in the communities concerned.

In some cases, such as that of the Yangtze in China, however, the benefits of massive infrastructure investment for flood control can be determined with some certainty and can be measured against the social costs, which in that case included massive relocation of people; while the benefits of generating renewable energy from hydropower are visible at a global scale through their contribution to reducing carbon dioxide emissions. This highlights the fact that one outcome of good water management, whether in the lowlands of Bangladesh or around Japan's Lake Biwa, may simply be that social and economic life is not disrupted as often as it might otherwise have been by droughts and floods; in the case of Japan, the number of days of inundation or of water restrictions has been recorded for decades.

More difficult to assess are those cases where better water management involves tradeoffs between different social objectives. Whether in the Snake River Basin in the affluent USA or the rapidly growing economies of the Mekong Basin, decisions made about the balance sought between livelihoods, energy production and environmental protection should reflect local priorities, which are inherently subjective. All that can be determined in such cases is whether the balance between interests appears to be reasonable. One measure of this, as in the management of groundwater in Denmark, is that both the water users and those who were impacting the quality of the resource are willing to pay the additional costs of protecting it.

But an important message is that integrated water resources management is not always a zero sum game. While, often, the challenge is to achieve a balance of interests and tradeoffs between them, the approach also offers opportunities to identify and support win–win synergies. This happened in India, where the incomes of Sukhomajri villagers rose even as water quality improved for their downstream neighbours; in Bangladesh, where the restoration of wetland ecosystems has had a positive impact on the communities dependent on them; in South Africa, where mining companies helped municipalities to pay the cost of treating the community's wastewater so that they could use it for their extractive processes; and it has also happened in the Lerma–Chapala Basin in Mexico, where effective control of water use enabled the environment to be protected even as local communities and their economies grew.

So, very importantly, there is not necessarily a contradiction between environment and development. In many of the cases presented, engagement in better water management has both improved the environment and brought social and economic benefits.

The processes described have occurred over many years – often reflecting but sometimes also leading societies' priorities. This is because the priorities for water management will often change, reflecting the political priorities of the times. So the Mekong River Commission's relatively limited impact can be understood in terms of the contested politics of the region over the past 50 years. Meanwhile, in South Africa, many of the water reforms were the result of majority interests finally holding sway over those of the formerly dominant minority. And, in China, developments of hydropower, transport, agriculture and water transfer all served vital strategic elements of national development policy. In so doing, however, they have raised questions about environmental protection

and social justice that have, in turn, contributed to significant changes in national policies.

So, in the policy sense, water may also be a lead sector, pioneering new issues and new approaches to governance as development challenges become more complex. Precisely because it is a complex domain to govern and manage, resistant to the simplistic administrative divisions that serve in other areas of human activity, the water resource sector does offer different models of governance that may be adopted elsewhere.

Distilling the key messages

By analysing the common threads that ran through many of our case studies, we have been able to distil five principal messages in each of four key categories: the objectives and outcomes of good water management; what good water resources management entails; how the management of water differs at different scales; and the nature of the IWRM approach itself.

Our five principal messages on the objectives and outcomes of good water management are:

1   Societies will use their own practices of governance to determine the appropriate balance between social, economic and environmental goals, which will change over time.
2   An important outcome of good water management is often that social and economic life is more secure than it would otherwise have been.
3   The most important determinants, as well as outcomes, of better water management will usually be found outside the water sector.
4   'Optimizing' economic growth, social equity and environmental sustainability implies that there will be compromises and tradeoffs, and that there are no unique 'best' solutions.
5   There is not *necessarily* a contradiction between the protection of the environment

and promotion of economic and social development.

Likewise, our five key messages on what good water resources management entails are:

1 Managing water effectively requires the sustained collective effort and engagement of women and men in all sectors of society if it is to be successful in achieving the society's goals.
2 Good water management needs robust, competent and trusted institutions, as well as economically, socially and environmentally sound investments in infrastructure.
3 Policy reforms and their implementation will only succeed if underpinned by a sound technical foundation.
4 Pragmatic, sensibly sequenced institutional approaches that respond to contextual realities have the greatest chance of working in practice.
5 Advances in water resources management can help in pioneering approaches to governance that can be applied in other sectors that face complex development challenges.

Our examination of how the management of water differs at different scales yields the following five messages:

1 Better management of water at a local level often needs the support of a sound policy framework at regional and national levels.
2 In large river basins, effective governance from local to basin levels is a major challenge, requiring functions to be placed at appropriate levels.
3 While a basin perspective is vital in many contexts, it must be supplemented by overarching national policies if water management is to be effective.
4 Transnational governance is a special case requiring specific approaches.
5 Water resources planning and management must be linked to a country's overall sustainable

development strategy and public administration framework.

And finally, we offer the following five conclusions on the nature of the IWRM approach itself:

1 There is no 'magic bullet' for all situations. IWRM is an approach rather than a method or a prescription.
2 Although it usually involves tradeoffs, integrated water resources management need not be a zero sum game, and 'win–win' outcomes can often be found.
3 Successful IWRM efforts adopt an integrated approach in order to address specific development problems; they never have an integrated approach as their principal objective.
4 The IWRM concept reflects good practice rather than radical new directions; in fact, there are many excellent examples of IWRM in practice that pre-date the formal adoption of the concept in 1992.
5 The process of water management does not have an end point and will continually have to respond to new challenges and opportunities.

Using these conclusions as a starting point, we offer some further reflections on the IWRM approach and its application in practice in the sections that follow.

Evolution of the integrated water resources management approach

One of the features of almost all the cases presented is that they were not considered to be explicit applications of the integrated water resources management approach. Most of them indeed began before the concept was formalized (as in India and Chile, Japan, Mexico and China), while in others it may have been mentioned in passing (as in South Africa and Australia), but the actions described were not initiated as formal attempts to introduce IWRM and were not explicitly guided by the concept.

Yet, in all the cases described, in responding to the very different challenges faced at the different scales, a very similar basic approach was applied, which recognized:

- the unitary nature of the water resource;
- the physical interventions that could be adopted to manage it;
- the limits to those physical interventions; and
- the need for an institutional framework that:
  - brought stakeholders together in an equitable manner that gave voice to the weak as well as to the powerful;
  - sought to achieve a balance of interests between them and, within this, recognized the value and importance of the waters concerned;
  - identified the environmental dimension of water management either explicitly as a separate 'use' or as a desirable outcome; and
  - developed organizations able to promote the overall approach.

It is important to note this because IWRM has been attacked as an unrealistic approach. The authors would agree that, if IWRM is seen as a fixed prescription, requiring the deployment of all the tools that are available in its armoury, it is not particularly helpful, if only because of the confusion that would ensue. Likewise, where there has been a focus on individual tools, success has also been limited. The textbook tradable water rights introduced in Pinochet's Chile may have achieved greater efficiency and productivity through reallocation between economic sectors, but it failed to address key social and environmental concerns and had to be substantially revised. Similarly, the establishment of river basin organizations, often taken as a doctrinaire first step, has played only a subsidiary role in improving water management in many countries (e.g. South Africa and Chile) and no

role at all in others (Denmark and Japan). Applying context-appropriate instruments in an appropriate sequence is certainly more important than any particular instrument in itself.

The point is that IWRM is *not* a prescription. Rather, it is an approach that offers a framework within which the problems of different communities and nations can be addressed.

That is not necessarily how the critics see it however, as a brief review reveals.

Watson, Walker and Medd (2007), writing in a special issue of *The Geographical Journal* devoted to what the journal describes as Integrated Water Management (IWM), note that there is 'very little agreement regarding what IWM means in practice', and that 'it is not altogether clear at the present time precisely what kinds of competencies, in the form of institutional and policy approaches, are required in order to successfully implement IWM in the "messy" and "turbulent" conditions that are increasingly evident in river basin systems around the world'.

A stronger challenge has come from Biswas (2004: 248), who notes that 'while at a first glance, the concept of IWRM looks attractive, a deeper analysis brings out many problems, both in concept and implementation, especially for meso- to macro-scale projects', arguing that this concept has 'a dubious record in terms of its implementation, which has never been objectively, comprehensively, and critically assessed'. Biswas and Tortajada (2004: 249) examine the broader issue of the implementation of the sustainable development concept, and its potential application to make water management more efficient and equitable than at present. They raise questions about whether 'a single paradigm of sustainable water resources management can encompass all countries of a very heterogeneous world, with very different cultures, social norms, climatic conditions, physical attributes, management and technical capacities, institutional and legal frameworks, and systems of governance'.

More recently, an article by Biswas (2008), entitled 'Integrated Water Resources Management:

Is It Working?', argues that the results of IWRM application:

> in a real world to improve water policy, programme and projects at macro- and meso-scales have left much to be desired. At a scale of 1 to 100 (1 being no integrated water resources management and 100 being full integration), any objective analyst will be hard-pressed to give a score of 30 to any one activity anywhere in the world in terms of its application.

In a response to one of these articles, Dukhovny (2004) notes that some of the criticism of IWRM is not a criticism of the IWRM concept itself, but rather of 'the flexible use of terminology as a general slogan, which is used very often because it is fashionable'. And, in some respects, it does appear that the diagnosis of failure is based on the examination of a straw man. Critically, the integration that Biswas finds missing is not that originally outlined at the Earth Summit in Rio de Janeiro in 1992, which recommended simply that:

> 18.16. Water resources development and management should be planned in an integrated manner, taking into account long-term planning needs as well as those with narrower horizons, that is to say, they should incorporate environmental, economic and social considerations based on the principle of sustainability; include the requirements of all users as well as those relating to the prevention and mitigation of water-related hazards; and constitute an integral part of the socioeconomic development planning process.

A reading of the entire Chapter 18 of Agenda 21, which addresses water issues, reveals that the key elements identified for integration are relatively straightforward. They include:

- the components of the resource (principally surface and groundwater);
- the quantity and quality of the resource;
- water, as part of ecosystems;
- users and the uses they make of water;
- water management and management of land uses; and

- water management and national development planning.

These dimensions of integration can be encapsulated in two simple concepts. The first is that water is a unitary resource despite the fact that its different elements often appear to be unrelated. The other is that water uses must be managed in a coordinated way. While their management is often deeply embedded in different areas of activity, such as agriculture, human settlements and industry, the water use in those individual sectors cannot in the long run be managed independently of the others; further, the activities of upstream and downstream users are inextricably interrelated.

This is not a particularly surprising finding. It is unusual to find any country that is able to manage its water sustainably while ignoring the links between underground and surface waters (although there are many countries where one or the other is not particularly important). Similarly, the obvious ability of water flows to dilute pollution (or to concentrate it, in times of drought) has usually convinced policy-makers that water quality cannot usefully be managed in isolation from its quantity. And all but the most blind of regimes recognizes that it is desirable that hydropower reservoirs should be managed not just to maximize returns for their owners and operators, but also in a way that protects downstream communities from floods (it is relevant that an unusual example of the contrary practice was in newly independent Mozambique where the managers of the Cahora Bassa reservoir on the Zambezi, the remnants of an old colonial regime desperately seeking to amortize its investments, unleashed devastating floods on the river valley in 1977).

Thus, on the basis of Agenda 21's simple and robust definition of integration, many countries would score relatively highly on the application of an IWRM approach, in terms of policy if not always in practice.

Many challenges remain to the application of IWRM in practice

This is not to say that challenges do not remain. There are still many examples of countries that miss opportunities, compound problems or create new ones because they do *not* manage water in an integrated way.

From the perspective of this book, perhaps the most important theoretical works in recent years are those that analyse the difficulties of applying IWRM approaches in practice. One example of this literature is Jeffrey and Kabat's (2003) *Integrated Water Resources Management: A post-natal examination*, which tries to assess the extent to which IWRM theory has been translated into practice and to identify research gaps. The paper notes that 'the concept possesses two major weaknesses from which the bulk of its perceived failings arise: the nature of the science which has informed its development, and its curiously ambiguous character in terms of contemporary intellectual paradigms'.

Without underplaying the importance of theoretical analysis, the fact remains that public administration is more an art than a science and that policy implementation necessarily occurs in the fuzzy context of real political life. The experience of public administration schizophrenically incorporating the conflicting approaches of Weberian bureaucracies and those of the so-called New Public Management philosophy is not by any means limited to water management. And the successes and failures of water resource management should be assessed in this practical context.

These issues become clear in a further set of papers put forward by researchers at IWMI and their partners, which focus on implementation of better management practices in developing country and 'informal economy' contexts. Shah and von Koppen (2006), in 'Is India Ripe for Integrated Water Resources Management?', raise serious questions about the IWRM paradigm with respect to India, including whether implementing IWRM is feasible in India in today's context and whether implementing IWRM has helped counter water scarcity and poverty in other countries with similarly densely populated basins, where the vast bulk of water users have few links to formal institutions.

Saravanan (2006), in a response to the IWMI article, argued that 'IWRM in its current package has been dismal worldwide, that formal and informal mechanisms both have their advantages and disadvantages, and it is the combination of these that contributes towards integrated water resources management'.

Indeed, it is in India that examples such as the Sukhomajri project demonstrate that IWRM approaches can work well at a local level; it is at the larger scale that the challenges of organizing collective action become more acute. It is at the level of relations between India's federal states that the consequences of the failure to establish robust institutions are most evident, a failure whose impact is not limited to the management of water resources.

The challenge of managing water at the macro-level is highlighted by the Comprehensive Assessment of Water Management in Agriculture (2008), which notes that, in developing countries, 'what usually gets passed-off in the name of IWRM at the operational level takes a rather narrow view of the philosophy' and has tended to focus on a blueprint package including national water policy, a water law and regulatory framework, recognition of the river basin as the unit of planning and management, treating water as an economic good, and participatory management. The report notes that this package represents a 'significant shift from current paradigms', that 'making this transition is proving to be difficult', and that consequently 'the so-called IWRM initiatives in developing country contexts have proved to be ineffective at best and counterproductive at worst'.

Since the Global Water Partnership was established in the mid-1990s, it has sought to explain and further develop the concept of IWRM in a series of publications. These included the GWP Technical Committee (TEC) Background Paper series, which addressed key conceptual issues related to water resources management; the Catalyzing Change series, which is designed to support countries in their practical efforts to improve water management; and the reports, case studies and publications associated with the GWP ToolBox. (A full list of relevant publications by the Global Water Partnership is included at the end of this chapter.)

So, if there has been a failure, it has been in the attempts to fit the IWRM approach that is observed in the messy world of practice into more tidy theoretical paradigms. There is clear tension between the world of practice and the attempts to design a theoretical framework that can encompass the practice of an extraordinarily diverse and complex sector.

This is one reason why, more recently, GWP has returned to the recommendations of the 1992 Earth Summit and sought to encourage practitioners to locate IWRM initiatives more explicitly within the framework of national development policies and strategies, both to ensure that they contribute to the achievement of national goals and also to ensure that the approaches adopted are coherent with the broader political economy of the society concerned (GWP Technical Committee, 2008). One practical outcome of this in South Africa was that local government legislation reflected the water services legislation that had preceded it; similarly, local Danish and Japanese water regulations preceded and guided national and regional environmental legislation.

So the case studies in this book have provided an opportunity to learn from the practice of good water management, rather than to build ever more complex theoretical frameworks.

This approach takes us back to the issues outlined in the introduction. It enables us to learn from a history that offers many examples of careful and sustainable water management, which included many of the dimensions of today's IWRM – ranging from the great irrigation works of Dujiangyan built two millennia ago in China and still serving millions of farmers, to the developments of the Tennessee Valley Authority in the USA in the 1930s. But to understand the disputes about the nature of IWRM and its efficacy, it is helpful to consider some of the drivers that gave origin to the concept.

One was technical: the growing capability of systems analysis, made possible by the evolution of computing power, in the 1960s. This saw water resource planners able, for the first time, to build complex models of hydrological systems, as illustrated by the Harvard Water Program. This programme, which ran from 1955 to 1960, was a large-scale interdisciplinary research endeavour that focused on developing planning and design methodologies for complex multi-purpose water-resource systems at the river-basin level. In particular, the publication of *Design of Water Resource Systems* (Maass et al) in 1962 had a huge impact on thinking among water resource specialists and led to several large-scale efforts to apply the approach in different contexts, with varying results.

The models developed as a result of these efforts enabled planners to go beyond hydrology and to combine economic and engineering analysis, drawing on techniques of mathematical efficiency models and computer simulation for river systems. One example of this was a study of the Rio Colorado basin in Argentina that applied multi-objective investment criteria and mathematical modelling for the generation of alternative development plans for consideration by decision-makers (Major and Lenton, 1979). It is the kind of analysis that informed the planning of development in complex river basins such as the Yangtze, since it also offered economic models to support the political processes of decision-making, and that still holds powerful sway today – much of the work of the Mekong River Commission has been devoted to developing such models.

A second driver was the growing understanding of the interaction between land use and rainfall and the impacts of different land uses on the infiltration of water into, and its flow over, the land. The notion of managing water and land together has been around since at least the beginning of the last century, when the need for comprehensive development of land, water and other natural resources first began to be felt (Watson, Walker and Medd, 2007). Further systematic work on land and water resources management contributed to the articulation of the principles of IWRM. Lundqvist, Lohm and Falkenmark (1985), for example, discussed strategies for river basin management with a particular focus on the integration of land and water in a river basin, highlighting the need to manage land and water resources in a coordinated way. This kind of thinking certainly influenced the approach taken to protecting local water resources in Denmark, where the interaction between land use and water quality was particularly acute.

However, the real catalyst that unleashed the concept was the growing awareness of the impact of human activities on the natural environment and the need for human societies to live within their environmental means. This matured in the late 1980s and 1990s in the events leading up to and following the landmark Earth Summit in Rio de Janeiro in 1992. Much of the thinking and action around water management issues at that time was inspired by the publication in 1987 of the 'Bruntland Report' (World Commission on Environment and Development, 1987), with its inspiring call for sustainable development as 'development that meets the needs of the present without compromising the ability of future generations to meet their own needs'. And, together with preparatory work in the water sector such as the 1992 International Conference on Water and the Environment (ICWE) in Dublin, Ireland, this is what triggered the formal adoption of integrated water resources management as the approach that offered the best chance to secure the world's water future.

At the Earth Summit, integrated water resources development and management was formally placed on the international agenda, the separate terms used to emphasize that even as the importance of the 'soft tools' of *management* was recognized, there was still a need for the 'hard tools' offered by infrastructure *development*. And following Rio, the term IWRM became part of international sustainable development vocabulary, a position that was reinforced at the World Summit on Sustainable Development in Johannesburg in 2002. Through this period, the key advance was that environmental goals became, in themselves, important objectives of water resource management rather than, as before, incidental considerations.

And it was in the nature of the historic compromise reached at Rio between 'developed' countries and those still 'developing' to focus on 'sustainable development' – recognizing as it does the inextricable links between the human condition and the state of the natural environment – that the social implications of water use were also placed squarely on the agenda. If the environment was addressed through the term 'sustainable', the balance in global society between social equity and economic growth was contained in the word 'development'.

The ongoing challenges of participation and adaptation

There are perhaps two dimensions in which IWRM as a concept remains justifiably contentious. The first is in its linkages with concepts of New Public Management and the economic prescriptions of the Washington Consensus, which have emphasized the role of markets and performance-based institutions. This in many ways reflects the dominant ideology of the period in which IWRM emerged as the guiding concept for water management rather than being essential elements of IWRM itself.

It is thus relevant to note that most of the examples presented in these chapters are not of the application of market mechanisms to guide water management, although they may have played a subsidiary part. The market, as an instrument of allocation, is not central to the concept of IWRM. What is critical is that there should be mechanisms to allocate and govern the use of water that reflect the value of water in addressing the social and economic imperatives of the societies concerned, and that encourage the efficient use of water where it is used for economic purposes. And it is not assumed that private institutions should take preference over public ones.

If there is a dominant political theme in the IWRM concept, it is about democracy and the importance of devising mechanisms that enable the participation of all interested parties in timely decisions about water and its management. So Kidd and Shaw (2007) note that while 'Integrated Water Management' (IWM) is traditionally viewed as a water-centric concept, many of the major challenges for integration actually arise at the interfaces between water and other natural, sectoral, territorial and organizational systems. Medd and Marvin (2007) discuss the role that intermediary organizations play in water governance processes, while Warner (2007) examines multi-stakeholder platforms and their effectiveness and sustainability, highlighting the difficulties of these platforms in dealing with integrated water management.

Some interest groups have promoted an extreme form of direct participation, in certain cases participation deliberately designed to block any human activity that impacts the environment. But the cases presented in this book demonstrate that what matters is that the different voices should be heard and heeded in water management. The institutional arrangements to achieve this may be different and may be more direct in smaller communities. But, particularly at the larger scales, IWRM does not prescribe a replacement of institutions of democratic governance with special mechanisms for water. It simply highlights the importance of that participation and the need to ensure that the boundaries of water and governance are aligned as far as possible (which does, in some cases, give rise to particular governance arrangements for water, often at basin scale, but as a practical requirement not an ideological prescription). And it does not suggest that local

interests should be able to override larger community interests.

Governance and participation issues are also important in the context of trans-boundary rivers, where a key question is the extent to which the powers of national organs can be taken over by a regional river basin organization. This challenge is highlighted in the Mekong case where, despite decades of cooperation in the Mekong Commission, decisions on future hydropower development are now being taken by national governments with only limited involvement of their neighbours. It remains a moot point whether this represents a failure of IWRM or whether the collaborative approach adopted in the Basin Commission has informed the local (in this case national) decisions that are now being taken, since appropriate subsidiarity of decision making is an important principle of IWRM.

These issues of individual interests and collective action have been addressed in depth by Kerr (2007) on the basis of field experience in India. He comments that:

> Theories from commons research predict great difficulty in managing complex watersheds and explain why success has been limited to isolated, actively facilitated microwatershed projects with a focus on social organization. Encouraging collective action is easiest at the microwatershed level but optimal hydrological management requires working at the macrowatershed level. Research suggests potentially severe tradeoffs between these two approaches. Resolving the tradeoffs is necessary for widespread success in watershed development but solutions are not clear.

It is precisely in this difficult terrain that the IWRM approach offers practical guidance, providing a basis to guide the choice of instruments appropriate to the particular scale.

IWRM as adaptive management

Perhaps the most important future challenge for IWRM as a management approach centres around its ability to address the test posed by global climate change. Questions have been raised about the ability of IWRM approaches to address the challenges of uncertainty and variability and, within these, the rapid changes forecast to take place as a consequence of climate change.

Practising water managers are sometimes puzzled by these concerns, which seem to indicate a lack of understanding of their terrain by the more theoretical commentators. Water resource management has traditionally been about understanding and quantifying risk and uncertainty, and designing robust responses to it. An interesting indicator of this is the extent to which engineers and hydrologists are being recruited to work in the financial markets precisely because of this ability to conceptualize and quantify risk in complex situations (Creamer Media Reporter, 2008).

However, a useful focus has emerged on the concept of 'Adaptive Water Management', as advanced by Pahl-Wostl et al (2005), Pahl-Wostl and Sendzimir (2005), and Timmerman, Pahl-Wostl and Moltgen (2008). Pahl-Wostl et al (2005) describe adaptive management as 'seeking to increase the adaptive capacity of river basins based on an understanding of key factors that determine a basin's vulnerability'. Pahl-Wostl and Sendzimir (2005) discuss the relationship between IWRM and Adaptive Water Management; the paper contains a summary of traditional water management and introduces IWRM as a response to overcome shortcomings of these regimes, and adaptive management as a concept that has its origins in ecosystem management. The paper highlights differences that differentiate traditional and integrated and adaptive approaches (Table 14.1).

More recently, Timmerman, Pahl-Wostl and Moltgen (2008) have provided a comprehensive overview of the state of the art in European research on integrated water resources management on the

Table 14.1 *Differences between traditional and adaptive regimes in water resources management*

	Prevailing regime	Integrated, adaptive regime
Governance	Centralized, hierarchical, narrow stakeholder participation	Polycentric, horizontal, broad stakeholder participation
Sectoral integration	Sectors separately analysed, resulting in policy conflicts and emergent chronic problems	Cross-sectoral analysis identifies emergent problems and integrates policy implementation
Scale of analysis and operation	Transboundary problems emerge when river sub-basins are the exclusive scale of analysis and management	Transboundary issues addressed by multiple scales of analysis and management
Information management	Understanding fragmented by gaps and lack of integration of information sources that are proprietary	Comprehensive understanding achieved by open, shared information sources that fill gaps and facilitate integration
Infrastructure	Massive, centralized infrastructure, single sources of design, power delivery	Appropriate scale, decentralized, diverse sources of design, power delivery
Finances and risk	Financial resources concentrated in structural protection (sunk costs)	Financial resources diversified using a broad set of private and public financial instruments

Source: Pahl-Wostl and Sendzimir (2005)

topics of participation, transboundary regimes, economics, vulnerability, climate change, advanced monitoring, spatial planning and the social dimensions of water management, and promote the concept of adaptive water management as the preferred direction for the future development of IWRM.

While this publication suggests that IWRM is inherently adaptive, other commentators still raise doubts, such as Medema and Jeffery (2005):

> … strategic policy in the water sector has developed from supply oriented, through demand oriented to integrated approaches over the past decades. Although IWRM as an abstract model has been widely accepted as the appropriate framework to deal with complex water resources management issues, its principles have not been elaborating on management under uncertainty, nor did they specifically articulate adaptive capacity as a significant feature of water management strategies. AM (adaptive management), as a concept, has been designed to support managers in dealing with uncertainties. In relation to the water sector, AM is considered an approach that could improve the conceptual and methodological base and promote realization of the goals of IWRM.
>
> (Medema and Jeffery, 2005)

Some practitioners in countries where an IWRM approach has been adopted believe that the way it is applied – in particular, the facility for on-going engagement between users – provides a framework within which adaptation can occur. In many of the cases presented, the IWRM approach has already demonstrably served to facilitate the adjustment to changing social priorities and economic circumstances.

In the South African case, for instance, the provision in national policy and legislation for a regular five-year review of resource management strategy is explicitly designed to cater for changes in social priorities as well as in the resource itself. Water-sharing treaties with neighbouring states, developed within this framework, outline arrangements for 'normal' conditions, but also provide a procedure to be adopted when, as during a drought for instance, the 'normal' can no longer be applied.

The Angas-Bremer case from Australia is another example of a management approach that has emerged, in part, to deal with changes in the resource as well as changes in the social and economic use of the resource, an approach that was put to practical test by climatic extremes during the recent drought.

These examples highlight that the process of water management does not have an end point unless the society itself is static and the cycle that drives water resource availability is not changing. Indeed, the management of water resources must be a pre-eminent example of how to address the broader social challenge of managing the various natural 'commons', on which all societies depend, in the face of changing natural and social circumstances.

In fact, it is arguably in the context of climate change and related challenges that integrated water resources management is an idea whose time has come. Mitigating the effects of climate change through greater reliance on renewable energy will require difficult and often contentious decisions and tradeoffs in the water resource management arena, such as tapping the world's hydroelectric energy potential in ways that are socially and environmentally responsible and allocating scarce land and water resources to the increased demand for biofuel production. Adapting to climate change and its resulting increases in climate variability (e.g. variations in the amount of rainfall or an intensification of the frequency and severity of storms) will require countries to develop more effective ways to deal with floods and droughts and their huge impacts on the economy.

Adaptation will also require greater flexibility to respond to changes in water supply and demand. The increasing risks posed by floods and cyclone-induced storm surges to people living in low-lying coastal areas in countries like Bangladesh will demand urgent attention, guided by robust strategies designed to address future trends not just immediate challenges.

More generally, efforts to meet the Millennium Development Goals (MDG) – the time-bound goals and targets to reduce poverty in all its manifestations that were set by the world's leaders in the year 2000 – highlight the water challenges that will be faced. Although the deadline of 2015 is unlikely to be met, meeting the first goal, to reduce income poverty and hunger by half across the developing world, will require that more water be used, more effectively, to support rural livelihoods and to grow more food. Yet this comes at a time when withdrawal and consumption of water for other purposes is increasing and is already causing significant environmental stresses in many river basins around the world.

Addressing the MDG to improve environmental sustainability will require sound management of freshwater resources to take full advantage of the services that nature currently provides – such as controlling floods and pollution through wetlands and river-floodplains. And achieving the goals of improving health, reducing childhood mortality, and furthering equality between women and men, cannot occur without more reliable and readily accessible supplies of unpolluted water as well as sanitation services to the unserved poor.

It is in the face of these challenges that the value of the IWRM approach can be seen, as a tool to advance the wider development goals of the very different societies in which it is applied, rather than as an end in itself. This is why this book has not been primarily about water but rather about the ways in which the intelligent management of water is already helping to further human wellbeing, generate wealth, and sustain the natural environment in the face of the immense challenges of the 21st century.

It is why the final message of the book is to reinforce what was said in the introduction: the world need not face a water crisis if it manages its limited water resources wisely. The potential crisis is a real one, but it is a crisis of poor water governance not of water scarcity. And the structured application of intelligent water resource management approaches, embodying the simple but logical elements of IWRM, could go a long way to help averting it.

References

Biswas, A.K. (2004) 'Integrated water resources management: a reassessment', *Water International*, vol. 29, pp248–256

Biswas, A.K. (2008) 'Integrated Water Resources Is It Working?', *Water Resources Development*, vol. 24, issue 1, pp5–22

Biswas, A.K. and Tortajada, C. (eds) (2004) *Appraising the Concept of Sustainable Development: Water Management and Related Environmental Challenges*, Oxford University Press, Oxford

Comprehensive Assessment of Water Management in Agriculture (2008) *Developing and Managing River Basins: The Need for Adaptive, Multilevel, Collaborative Institutional Arrangements*, issue Brief No.12, International Water Management Institute and Global Water Partnership, Colombo, Sri Lanka. Comprehensive Assessment Secritariat

Creamer Media Reporter (2008) 'Investment banker may hire engineers to strengthen financial skills', *Engineering News, South Africa* (8 July 2008)

Dukhovny, W. (2004) Discussion of 'Biswas, A.K. (2004) "Integrated Water Resources Management: A Reassessment". *Water International* 29(2), 248–256' in: Discussion Notes: *Water International*, vol. 29, issue 4, pp530–535

Global Water Partnership Technical Committee (2008) *How to integrate IWRM and national development plans and strategies and why this needs to be done in the era of aid effectiveness*, Catalyzing Change Series Policy Brief 6, Global Water Partnership, Stockholm

Jeffrey, P. and Kabat, P. (2003) *Integrated Water Resources Management: A post-natal examination*, contributed paper to the discussion about Methods for Integrated Water Resources Management (MIWRM) and Transboundary Issues as a part of EC FP6 programme topic II.3.1 – Integrated water management at catchment scale

Kerr, J. (2007) 'Watershed Management: Lessons from Common Property Theory', *International Journal of the Commons*, vol. 1, issue 1, pp89–109

Kidd, S. and Shaw, D. (2007) 'Integrated water resource management and institutional integration: realising the potential of spatial planning in England', *The Geographical Journal*, vol. 173, issue 4, pp312–329

Lundqvist, J., Lohm, U. and Falkenmark, M. (eds) (1985) *Strategies for River Basin Management, Environmental Integration of Land and Water in a River Basin*, D. Reidel Publishing, D. Reidel Publishing Co., Dordrecht, The Netherlands

Maass, A., Hufschmidt, M.M., Dorfman, R., Thomas, Jr., H.A., Marglin, S.A. and Fair, G.M. (1962) *Design of Water-Resource Systems; New Techniques for Relating Economic Objectives, Engineering Analysis, and Governmental Planning*, Harvard University Press, Cambridge, MA

Major, D.C. and Lenton, R.L. (1979) *Applied Water Resource Systems Planning*, Prentice Hall, Englewood Cliffs, NJ

Medd, W. and Marvin, S. (2007) 'Strategic intermediation: between regional strategy and local practice', *Sustainable Development*, vol. 15, pp318–327

Medema, W. and Jeffrey, P. (2005) *IWRM and Adaptive Management: Synergy or Conflict?* NeWater, Deliverable 1.1.1

Pahl-Wostl, C., Downing, T., Kabat, P., Magnuszewski, P., Meigh, J., Schüter, M., Sendzimir, J. and Werners, S. (2005) *Transition To Adaptive Water Management: The NeWater project*, NeWater Working Paper 1, Institute of Environmental Systems Research, University of Osnabrück

Pahl-Wostl, C. and Sendzimir, J. (2005) *The Relationship between IWRM and Adaptive Water Management*, NeWater Working Paper 3, Institute of Environmental Systems Research, University of Osnabrück

Saravanan, V.S. (2006) 'Integrated Water Resource Management: A Response', *Economic and Political Weekly*, vol. 41, issue 38, pp4086–4087

Shah, T. and von Koppen, B. (2006) 'Is India Ripe for Integrated Water Resources Management: Fitting Water Policy to National Development Context', *Economic and Political Weekly*, vol. 41, issue 31, pp3413–3421

Timmerman, J.G., Pahl-Wostl, C. and Moltgen, J. (2008) *The Adaptiveness of IWRM: Analysing European IWRM Research*, London: IWA Publishing

Warner, J. (2007) (ed.) *Multi-stakeholder Platforms for Integrated Water Management Studies in Environmental Policy and Practice*, Ashgate Publishing, Aldershot

Watson, N., Walker, G. and Medd, W. (2007) 'Critical perspectives on integrated water management', *The Geographical Journal*, vol. 173, issue 4, pp297–299

World Commission on Environment and Development (1987) *Our Common Future* (Brundtland report), Oxford University Press, Oxford

Additional references

Global Water Partnership (2003) *Integrated Water Resources Management Toolbox, Version 2*, GWP Secretariat, Stockholm

Global Water Partnership (2005) 'Resource links for developing IWRM and water efficiency strategies – useful links for developing an IWRM and water efficiency strategy', available from www.gwpforum.org/gwp/library/CatalyzingChange-Resources.pdf (accessed 29 May 2008)

Global Water Partnership (2006) *The Boldness of Small Steps* Global Water Partnership, Stockholm

Global Water Partnership Technical Advisory Committee (2000) *Integrated Water Resources Management*, TAC Background Papers No. 4, Global Water Partnership, Stockholm

Global Water Partnership Technical Committee (2003) *Poverty Reduction and IWRM*, TEC Background Papers No. 8, Global Water Partnership, Stockholm

Global Water Partnership Technical Committee (2004a) *Catalyzing Change: A handbook for developing integrated water resources management (IWRM) and water efficiency strategies*, Global Water Partnership, Stockholm

Global Water Partnership Technical Committee (2004b) *Unlocking the Door to Social Development and Economic Growth: how a more integrated approach to water can help*, Catalyzing Change Series Policy Brief 1, Global Water Partnership, Stockholm

Global Water Partnership Technical Committee (2006a) *Checklists for Change: Defining areas for action in an IWRM strategy or plan*, Catalyzing Change Series Technical Brief 1, Global Water Partnership, Stockholm

Global Water Partnership Technical Committee (2006b) *Gender Mainstreaming: An essential component of sustainable water management*, Catalyzing Change Series Policy Brief 3, Global Water Partnership, Stockholm

Global Water Partnership Technical Committee (2006c) *How IWRM will contribute to achieving the MDGs*, Catalyzing Change Series Policy Brief 4, Global Water Partnership, Stockholm

Global Water Partnership Technical Committee (2006d) *Mainstreaming Gender in Integrated Water Resources Management Strategies and Plans: Practical steps for practitioners*, Catalyzing Change Series Technical Brief 5, Global Water Partnership, Stockholm

Global Water Partnership Technical Committee (2006e) *Monitoring and Evaluation Indicators for IWRM Strategies and Plans*, Catalyzing Change Series Technical Brief 3, Global Water Partnership, Stockholm

Global Water Partnership Technical Committee (2006f) *Taking an Integrated Approach to Improving Water Efficiency*, Catalyzing Change Series Technical Brief 4, Global Water Partnership, Stockholm

Global Water Partnership Technical Committee (2006g) *Tools for keeping IWRM Strategic Planning on Track*, Catalyzing Change Series Technical Brief 2, Global Water Partnership, Stockholm

Global Water Partnership Technical Committee (2006h) *Water and Sustainable Development: Lessons from Chile*, Catalyzing Change Series Policy Brief 2, Global Water Partnership, Stockholm

Global Water Partnership Technical Committee (2007) *Climate Change Adaptation and Integrated Water Resources Management – An initial overview*, Catalyzing Change Series Policy Brief 5, Global Water Partnership, Stockholm

Index

Note: *an 'n' after a page number indicates a reference to a note.*